Pharmaceutical Product Development

DRUGS AND THE PHARMACEUTICAL SCIENCES

A Series of Textbooks and Monographs

Executive Editor

James Swarbrick

PharmaceuTech, Inc.
Pinehurst, North Carolina

Advisory Board

Pharmaceutical Product Development

In Vitro-In Vivo Correlation

edited by

Dakshina Murthy Chilukuri
U.S. Food and Drug Administration (FDA)
Silver Spring, Maryland, U.S.A.

Gangadhar Sunkara
Novartis Pharmaceuticals
East Hanover, New Jersey, U.S.A.

David Young
AGI Therapeutics, Inc.
Columbia, Maryland, U.S.A.

CRC Press
Taylor & Francis Group
Boca Raton London New York

CRC Press is an imprint of the
Taylor & Francis Group, an **Informa** business

First published 2007 by Informa Healthcare, Inc.

Published 2019 by CRC Press
Taylor & Francis Group
6000 Broken Sound Parkway NW, Suite 300
Boca Raton, FL 33487-2742

© 2007 by Taylor & Francis Group, LLC
CRC Press is an imprint of Taylor & Francis Group, an Informa business

First issued in paperback 2019

No claim to original U.S. Government works

ISBN 13: 978-0-367-45317-6 (pbk)
ISBN 13: 978-0-8493-3827-4 (hbk)

Visit the Taylor & Francis Web site at
http://www.taylorandfrancis.com

and the CRC Press Web site at
http://www.crcpress.com

Library of Congress Cataloging-in-Publication Data

Pharmaceutical product development / edited by Dakshina Murthy Chilukuri,
    Gangadhar Sunkara, David Young.
        p. ; cm. -- (Drugs and the pharmaceutical sciences ; v. 165)
    Includes bibliographical references and index.
    ISBN-13: 978-0-8493-3827-4 (hardcover : alk. paper)
    ISBN-10: 0-8493-3827-1 (hardcover : alk. paper)
    1. Drug development. 2. Drugs--Design. I. Chilukuri, Dakshina Murthy. II. Sunkara,
Gangadhar. III. Young, David, 1952- IV. Series.
    [DNLM: 1. Drug Design. 2. Biological Availability. 3. Drug Evaluation,
Preclinical--methods. W1 DR893B v.165 2007 / QV 744 P5348 2007]

RM301.25.P395 2007
615'.1--dc22   2006050423

---

*To my mother, Chilukuri Shyamalamba, my (late) father,
Chilukuri Ramachandra Murthy, and my wife, Ishwarya.*

Dakshina Murthy Chilukuri

*I dedicate this book to my father, Sri Prakasa Rao,
mother, Smt Ammajamma, wife, Nagalakshmi, and
son, Kiran Sai, for their loving support and caring.*

Gangadhar Sunkara

*Dedicated to my parents, Jack and Grace, my wonderful wife, Karen,
and my two fantastic children, Bethany and Kyle.*

David Young

Preface

The development of a new drug from the laboratory to the patient takes between 10 to 15 years and costs hundreds of millions of dollars. The pharmaceutical industry, in conjunction with the regulatory bodies across the world, has continuously sought to reduce the time to bring a new drug to the market and reduce the cost of drug development in order to maximize return on the investment and to bring drugs to patients sooner. In the last two decades, the pharmaceutical industry has experimented with and successfully adopted several integrated and multidisciplinary approaches to achieve these reductions in development efforts. These efforts were primarily made in the research areas of drug compound screening, toxicological evaluation, and pharmaceutical product development.

Owing to the scientific advances in the development of chemically complex therapeutic agents (e.g., recombinant proteins/peptides, gene-based drugs) intended for chronic therapies, there is a growing need for the continuous development of suitable formulation either as conventional immediate release oral formulation or controlled and/or continuous delivery via oral or parenteral routes. As formulation and delivery strategies can enhance competitive advantage for pharmaceutical companies, several chemical- and formulation-related issues are driving the early- and late-stage development of drug products. In the last few decades, significant medical advances have been made in the area of drug delivery with the development of novel dosage forms. However, the delivery of several classes of drugs continues to be a challenge, mainly due to short biological half-lives, poor membrane permeability, and associated toxicity in the administered doses. Since more is now known about the relationship between chemical properties and movement of drugs throughout the body, drug discovery scientists are able to consider the pharmacokinetic properties of agents much earlier in the drug/product development process.

Formulation development connects several key preclinical and clinical trials to support the new drug application. Formulation development and optimization involves varying excipient levels, processing methods, identifying discriminating dissolution methods, and subsequent scale-up of the final product. The quantitative and qualitative changes in a formulation may alter drug release and in vivo performance. Developing a pharmaceutical product formulation in a timely manner, while ensuring quality, is a complex process that requires a systematic, science-based approach. Thus, ever increasing pressures to reduce pharmaceutical product development timelines have resulted in pharmaceutical scientists in physical pharmacy, pharmaceutics, and pharmacology working collaboratively to develop an integrated approach of product development by addressing physicochemical and biological issues early on. Such continuous collaborative effort has resulted in the development of an important tool: in vitro–in vivo correlation.

It is expected that the readers of this book possess a basic knowledge of biopharmaceutics and pharmacokinetics. The material presented in this text serves the need of scientists who are keen on utilizing the principles of in vitro–in vivo correlation in drug development. The objectives of this book are three-fold: *i*) to serve as a useful tool to help guide scientists in research and development by outlining the theory and successful practice of in vitro–in vivo correlation, *ii*) to help formulators apply the tool in designing and developing prototypes that enable selection of clinical formulations, and *iii*) to help formulate strategy(ies) for product life-cycle management.

We would like to express our gratitude to the committed subject experts who have graciously agreed to spend their valuable time in writing the various chapters. Additional thanks to the publishers who have been supportive and considerate in an effort to bring out the best book possible.

Dakshina Murthy Chilukuri
Gangadhar Sunkara
David Young

Contents

Contributors

Sandhya K. Apparaju   Food and Drug Administration, Silver Spring, Maryland, U.S.A.

Ajay K. Banga   College of Pharmacy, Mercer University, Atlanta, Georgia, U.S.A.

Ayyappa Chaturvedula   Glaxo SmithKline, Parsippany, New Jersey, U.S.A.

Adrian Dunne   School of Mathematical Sciences, University College Dublin, Dublin, Ireland

Colm Farrell   Department of Pharmacokinetics and Biopharmaceutics, GloboMax—ICON Development Solutions, Marlow, Bucks, U.K.

Siobhan Hayes   Department of Pharmacokinetics and Biopharmaceutics, GloboMax—ICON Development Solutions, Marlow, Bucks, U.K.

Kevin C. Johnson   Intellipharm, LLC, Niantic, Connecticut, U.S.A.

Shoufeng Li   Pharmaceutical and Analytical Development, Novartis Pharmaceuticals Corporation, East Hanover, New Jersey, U.S.A.

Patrick J. Marroum   Office of Clinical Pharmacology and Biopharmaceutics, Center for Drug Evaluation and Research, Food and Drug Administration, Silver Spring, Maryland, U.S.A.

Nishit B. Modi   Department of Clinical Pharmacology, ALZA Corporation, Mountain View, California, U.S.A.

Martin Mueller-Zsigmondy   Pharmaceutical and Analytical Development, Novartis Pharmaceuticals AG, Basel, Switzerland

Srikanth C. Nallani   Food and Drug Administration, Silver Spring, Maryland, U.S.A.

Hequn Yin   Early Clinical Development, Novartis Pharmaceuticals Corporation, East Hanover, New Jersey, U.S.A.

David Young   AGI Therapeutics, Inc., Columbia, Maryland, U.S.A.

Dissolution: Fundamentals of In Vitro Release and the Biopharmaceutics Classification System

Kevin C. Johnson

Intellipharm, LLC, Niantic, Connecticut, U.S.A.

INTRODUCTION

Drug dissolution is nearly impossible to study in the environment that it is intended to occur, namely, the human gastrointestinal (GI) tract. As a surrogate, in vitro dissolution testing is required to ensure that drug dissolves at a consistent rate from batch to batch of formulated drug product. Beyond the important quality control function of dissolution testing, the goal of this chapter will be to show how dissolution in the context of drug absorption from the GI tract can be modeled to gain insight into the important factors that control the rate of dissolution, and as a result, provide a mechanistic basis for predicting a correlation between in vitro dissolution and in vivo time profile of drug in the blood. The potential benefit from this insight will arise from the recognition of the critical factors that must be understood and controlled, and how to best design tests to ensure the quality of the finished drug products. A mechanistic model also provides a more proactive approach to the design of dosage forms, increasing the chances that the dosage form will have the desired release characteristics resulting in an efficacious drug plasma profile.

We start with an oversimplified view of dosage forms by assuming that all drugs are given as solutions with no need for disintegration and dissolution testing. It assumes that only drug in solution can cross the GI membrane.

Given this scenario, drug discovery scientists would expect that the amount of drug available systemically would increase with increasing dose, allowing them to establish a safe and efficacious dose.

In trying to explain why the amount of absorbed drug would increase with dose, one would recall the principles of physical chemistry that state that matter will tend to move from a point of higher concentration or chemical potential; drug in solution within the GI tract, to a point of lower concentration; the systemic blood supply:

$$D_{lumen} \xrightarrow{K_a} D_{blood} \tag{1}$$

where D_{lumen} is the mass of drug dissolved in the GI lumen, K_a is a first-order absorption rate constant, and D_{blood} is the mass of drug absorbed into the systemic blood supply. For simplicity, metabolism has been ignored. A simple absorption rate equation used in pharmacokinetics will be useful in making the point:

$$\frac{dD_{lumen}}{dt} = -K_a D_{lumen} = \text{negative rate of drug absortpion} \tag{2}$$

Integrating this equation from dosing at zero time where $D_{lumen} = $ Dose to some later time yields:

$$D_{lumen} = \text{Dose} \times e^{-K_a t} \tag{3}$$

For solution doses, the direct proportionality can be seen between the dose and D_{lumen}. And because D_{lumen} is the driving force for absorption, increasing dose increases absorption. Again, this is based on the assumption that all doses are in solution. In the real world, this is not the case, and drug dissolution occurs at a finite rate and may not be complete due to inadequate solubility.

ABSORPTION

Absorption is important to discuss with regard to dissolution because in vivo, dissolution occurs in a permeable GI tract, whereas dissolution testing is usually done in an impermeable glass vessel. Characterizing absorption using parameters such as an absorption rate constant or permeability provide an essential link between dissolution and what happens to the drug once it has been absorbed, and necessary to establish a predictable in vitro/in vivo correlation. Absorption is an area of common interest to both formulation and pharmacokinetic scientists because it affects the decisions that both groups make.

In Equations 2 and 3, drug absorption was characterized as drug leaving the GI lumen. This is a common way to study absorption. For example, both rat and human intestinal perfusion experiments have been used to characterize absorption by isolating a segment of the intestine and using the difference in drug concentration entering and leaving the segment to calculate an absorption rate constant or permeability. When done properly to ensure that drug degradation or water absorption and secretion are not affecting the results, characterizing absorption in this way is a reliable method.

Another method to study absorption would be to use drug blood concentrations after intravenous and oral dosing to calculate an absorption rate constant. This method is more complicated because metabolism must be taken into account. For example, it would be possible for a drug to be completely absorbed across the GI membrane, but entirely metabolized by the liver before reaching the systemic circulation. Unless all drug metabolites were traced, one might erroneously assume that because the drug itself was not detected in the blood, that absorption had not occurred. This situation is important to recognize so that a formulation group does not waste time attempting to improve absorption when metabolism is the real problem.

Throughout this chapter, both the term absorption rate constant and permeability will be used interchangeably. The term permeability has the advantage that it is shorter and perhaps more descriptive. Both absorption rate constants and permeability are not like other physical parameters that might be found in the scientific literature. Their values have a rather large degree of error typical of pharmacokinetic parameters and may vary in the GI tract due to positional changes in anatomy and environmental conditions. The term absorption rate constant has been criticized because the name implies something that it is not. However, in practice, permeability is also usually treated as if it were a constant. The absorption rate constant has the advantage of having the characteristics of a first-order rate constant. Given its value, one can quickly take the natural logarithm of two and divide it by the absorption rate constant to calculate an absorption half-life. In effect, permeability is usually converted to a first-order rate constant for calculations that require the calculation of mass of drug absorbed.

There are several techniques to characterize permeability before going into human clinical studies, including rat intestinal perfusions, Caco-2 permeability (1), and parallel artificial membrane permeability assay (PAMPA) (2). However, one should be careful in considering whether the values are useful in an absolute sense or only in a relative sense. Clinically determined human absorption rate constants serve as a good reality check for comparison. There are a large number of clinical studies that have been done, which include some estimate of the absorption rate constant (3). The upper end of the range appears to be in the neighborhood of 0.1 min^{-1}, placing the shortest absorption half-life around seven minutes. Although there might be reports of individual patients having higher values, it would be rare that population averages exceed this value. Typically, if an intravenous dosing leg of a clinical study has been done in addition to an oral leg, then an observed absorption rate constant can be determined after correcting for bioavailability. As a word of caution, some reported absorption rate constants might be smaller than the true intrinsic absorption rate constants due to the influence of dosage form disintegration and drug dissolution. Correcting observed absorption rate constants for the effect of dissolution was part of the motivation for the author of this chapter in developing more sophisticated dissolution models so that surrogate methods could be validated against a more accurate data base of clinically determined absorption rate constants.

As mentioned earlier, one of the most direct methods for evaluating permeability is the isolated perfusion of the human intestine (4–7). Because some of this data will be used in this chapter, it is worthwhile to compare this method and its results with the traditional way of determining the absorption rate constant from pharmacokinetic data. From pharmacokinetics, the rate of absorption can be described by:

$$\frac{dM}{dt} = -K_a M \tag{4}$$

where M is the mass of drug in the GI tract. Similarly, the rate of absorption can be determined from an intestinal perfusion experiment:

$$\frac{dM}{dt} = -PAC \tag{5}$$

where P is the intestinal permeability, A is the surface area of the perfused intestinal segment, and C is the drug concentration. Because drug concentration is equal to the mass of drug in the segment divided by the volume V of the segment,

$$\frac{dM}{dt} = -PA\frac{M}{V} \tag{6}$$

Equating the pharmacokinetic rate of drug absorption with the rate determined from the perfusion method, the following relationship is derived:

$$K_a = P\frac{A}{V} \tag{7}$$

For the perfused segment, which is a cylindrical plug, the surface area divided by the volume of the plug is given by:

$$\frac{A}{V} = \frac{2\pi rL}{\pi r^2 L} = \frac{2}{r} \tag{8}$$

where r and L are the radius and the length of the perfused segment, respectively. Given an estimated radius of the human small intestine of 1.75 cm (7), the surface to volume ratio is approximately 1.1. Permeability is typically in units of cm/sec, whereas absorption rate constants are generally reported in units of reciprocal minutes or hours. Permeability can be easily converted to an absorption rate constant by multiplying its value by the surface to volume ratio and converting to desired units of time. Using propranolol as an example, its human intestinal permeability was reported as 3.878×10^{-4} cm/sec (6). Substituting this value into Equation 7 and converting the units for time from seconds to minutes gives:

$$K_a = 3.878 \times 10^{-4} \frac{cm}{sec} \times \frac{60\,sec}{min} \times 1.1\frac{cm^2}{cm^3} = 0.026\,min^{-1} \tag{9}$$

This calculated absorption rate constant using permeability from human intestinal perfusion experiments compares well with the value of 0.025 min^{-1} reported independently from a pharmacokinetic study (8).

It should also be noted that the absorption rate constant as presented earlier has only been shown in one direction: from the lumen to the blood. In general, drugs with solubilities in the mg/mL range will exist in the mg/mL range in the GI tract. Blood concentrations are generally in the μg/mL range. Therefore, the reverse absorption rate constant would have to be approximately 1000-fold higher to be significant. If the drug is poorly soluble, in the μg/mL range, blood concentrations are likely to be in the ng/mL range. Again, the reverse rate would have to be 1000-fold higher to be comparable to the forward rate. For the remainder of this chapter, the reverse rate will be ignored while acknowledging that this assumption is open to debate (9,10).

In reality, most dosage forms are tablets containing a crystalline powder of the drug substance. Unlike a solution dose, the amount of drug dissolved in the intestine will increase with time, as the dosage form disintegrates and releases crystalline drug particles. There are no simple pharmacokinetic equations to describe this process. Solution dosage forms of the same drug are unlikely to show differences in the rate and extent of absorption and, therefore, are likely to be bioequivalent. Solution dosage forms will present the total dose in the form of drug that can be absorbed (in solution) with the amount of drug in the lumen falling exponentially, as drug is absorbed at the same rate. However, immediate-release solid dosage forms are likely to have slightly different rates of disintegration due to the choice of tablet excipients and the manufacturing process and potentially larger differences in dissolution rate depending on the drug particle size and the efficiency of wetting provided by the formulation.

There is a mechanistically based theory to describe the kinetics of dissolution that will be discussed, and it will be shown how dissolution theory can be used to determine if the combined effect of disintegration and wetting are having a significant impact on drug absorption. Before getting into the more sophisticated treatment of dissolution, the absorption rate equation discussed earlier provides the starting point for a very simple and useful analysis of situations that might present difficulties in drug absorption. Recalling Equation 2, the integration of this equation over a specified period of time gives the mass of drug absorbed from the GI tract. If nothing limited the amount of drug that could be administered as a solution to the GI tract, then there would be no limit to the amount of drug that could be absorbed. However, drug solubility presents a limit to the amount of drug that can exist as a solution in the GI tract. Any solid crystalline drug administered would continue to dissolve unless its concentration equaled its solubility. At this point, no further drug would dissolve until some of the drug in solution was absorbed. If enough solid drugs were administered so that the rate of dissolution was equal to the rate of absorption, a temporary steady state would exist where the concentration of drug in the GI tract would be equal to the solubility of the drug. In this case:

$$D_{\text{lumen}} = \text{Sol}_{\text{gi}} \times V_{\text{gi}} \qquad (10)$$

Sol_{gi} is the solubility of the drug in the GI fluid, and V_{gi} is the volume of GI fluid present. Substituting this expression into Equation 2 yields:

$$\frac{dD_{\text{lumen}}}{dt} = -K_a \times \text{Sol}_{\text{gi}} \times V_{\text{gi}} = \text{negative rate of absorption} \qquad (11)$$

It can be seen that if enough solid crystalline drug is given so that the rate of dissolution can match the rate of absorption to keep the concentration of drug in the GI tract at its solubility, the rate of absorption becomes constant. If equation is integrated over the typical residence time, that drug would remain in the small intestine t_r, with this integration called the maximum absorbable dose (MAD) (11), a simple calculation is the result:

$$\text{MAD} = K_a \times \text{Sol}_{\text{gi}} \times V_{\text{gi}} \times t_r \qquad (12)$$

A dimensional analysis using typical units for the various constants shows that the MAD number would have the typical units of dose in milligrams:

$$\text{MAD} = K_a \left(\frac{1}{\text{min}}\right) \times \text{Sol}_{\text{gi}}\left(\frac{\text{mg}}{\text{ml}}\right) \times V_{\text{gi}}(\text{ml}) \times t_r(\text{min}) \qquad (13)$$

This provides a useful benchmark calculation that includes the key parameters that are generally recognized as limiting absorption: solubility in the GI tract and the intrinsic absorption rate constant specific to drug in solution. Other attractive features of the MAD number are that it is expressed in units of mass, facilitating communication among scientists with diverse backgrounds involved in pharmaceutical research, and the MAD number is dose-independent. This is particularly useful in early drug discovery and development because the clinical dose is unknown. The simplicity and pertinence of the MAD analysis in the drug discovery/early development phase has lead to its growing acceptance (12).

The other key parameter for absorption is solubility. Given everything else the same, dissolution rate will increase with solubility. Given two drugs with the same absorption rate constant, the one with the greater solubility will have a greater MAD. In measuring solubility, using a fluid that is closer to real GI fluid rather than plain water is likely to give a more accurate prediction of the MAD. Likewise, using a dissolution media that more closely mimics GI fluid is more likely to result in a meaningful in vitro/in vivo correlation between dissolution and absorption.

The MAD number is intended to give a "ballpark" estimate of how much drug one might expect to be absorbed if a plug of fluid with a volume expected to be found in the GI tract were to be saturated with drug, and that the drug in solution could exit the plug at a rate determined by the absorption rate constant for a period of time that the plug would typically reside in the small intestine. The typical fluid volume and GI residence time could evolve, as experience and data become available, but, for example, let them be 250 mL and three hours, respectively. In general, if the projected clinical dose were below the MAD number, then drug absorption should not be a limiting factor in determining clinical efficacy. However, if the projected clinical dose were above the MAD number, limited absorption would be likely. The same could be said for projected doses for toxicological studies, and the volume and residence time could be scaled to a particular animal.

The MAD number could also be used in early drug discovery to rank order candidates with regard to their ease of development. Given similar potency, a compound with a larger MAD number would have a greater dose/exposure range in which to establish safety and efficacy. Toxicity and clinical studies that show a plateau in exposure as a function of dose can be used to validate the predictive value of the MAD number.

Table 1 shows MAD calculations for several marketed drugs. An attempt was made to find literature values for solubility in the pH range of 6 to 7 to reflect conditions in the small intestine. Only one significant figure is shown for solubility, absorption rate constant, and MAD values due to the large degree of uncertainty associated with trying to assign numbers to parameters in an in vivo situation. Typical doses can also vary due to the size, age, sex, and genetics of the patient. However, inspection of Table 1 shows that for drugs that made it into the market as conventional products, namely atenolol, digoxin, furosemide, naproxen, and propranolol, the typical dose is below the MAD number that would be calculated based on the solubility of the drug at pH values expected to be found in the small intestine. For cyclosporine and griseofulvin, the dose is greater than the MAD number, and it is generally

Table 1 Maximum Absorbable Dose Estimates for Several Commercially Available Drugs

Drug	Solubility (mg/mL)	Peff (cm/sec)	Ka (1/min)	MAD (mg)	Dose (mg)
Atenolol	30	0.15×10^{-4}	0.001	1000	100
Carbamazepine	0.3	4.3×10^{-4}	0.03	400	400
Cyclosporin	0.03		0.02	30	300
Digoxin	0.05		0.05	100	<1
Furosemide	3	0.3×10^{-4}	0.002	300	80
Griseofulvin	0.008		0.1	40	500
Naproxen	1	8.0×10^{-4}	0.05	2000	500
Nifedipine	0.01		0.07	30	20
Propranolol	30	3.88×10^{-4}	0.03	40,000	160

Abbreviation: MAD, maximum absorbable dose.
Source: From Refs. 4, 5, 8, 13, 20, 26, 29–36.

Table 2 Percent of a Solution Dose Absorbed in Three
Hours for Various Absorption Rate Constants

Absorption rate constant (1/min)	Percent absorbed in 3 hr
0.03	100
0.01	83
0.003	42
0.001	16

known among formulation scientists that extensive work has been carried out on the development of dosage forms to improve the absorption of cyclosporin and griseofulvin. The absorption rate constant for griseofulvin was assumed to be at the high end of the range. Even so, the MAD number is less than the dose. This demonstrates that, in some cases, only solubility needs to be measured to determine a likely problem with absorption.

Both nifedipine and carbamazepine are borderline cases where the doses of the immediate-release dosage form are similar to the MAD. However, based on the commercially available dosage forms, the need for solubility-enhancing formulations to improve the bioavailability for nifedipine or carbamazepine does not appear as critical as for cyclosporine and griseofulvin. The intent of the MAD analysis summarized in Table 1 is to demonstrate that the degree of difficulty in developing a commercial dosage form with regard to absorption can be estimated in a relatively straightforward manner.

Inspection of Table 1 indicates that although atenolol has the lowest absorption rate constant, it has a high MAD number. Table 2 shows the predicted percent of dose absorbed for solution doses for various values for the absorption rate constant. For a drug with a low absorption rate constant like atenolol, nothing can be done to improve the percent of dose absorbed without altering the characteristics of the intestinal membrane. However, as long as solubility does not prevent the entire dose from dissolving, increasing the dose will continue to increase the absolute amount of drug absorbed, even while the percent of dose absorbed remains the same. While some may view incomplete absorption due to low permeability unfavorably, it does not present an obstacle to increasing absorption as long as solubility does not limit absorption.

DISSOLUTION AND ABSORPTION, DISTRIBUTION, METABOLISM, AND EXCRETION

Although the MAD analysis provides a simple and valuable approach to understand and act on solubility and permeability data, much more can be done with regard to modeling dissolution and absorption, and at the same time, incorporating pharmacokinetic concepts, such as metabolism, excretion, and distribution of drug in and out of tissues. By taking a more comprehensive approach to modeling the whole process, commonly referred to absorption, distribution, metabolism, and excretion (ADME), dissolution can be correlated to blood plasma concentrations and, therefore, C_{max} and area under concentration time curve (AUC).

In developing an in vitro/in vivo correlation, a mechanistically based approach will be described. This approach is distinct from perhaps the more common and traditional empirical approach. With the empirical approach, dosage forms are made with varying dissolution rates, the resulting dosage forms are dosed in the clinic to determine plasma concentrations, and finally, the plasma concentrations are correlated with the dissolution rates. This approach does not require a mechanistic explanation of the result. Its limitation is that it does not provide a mechanistic framework to predicting outcomes across chemical structures and, therefore, may not be applicable to the development of future drugs. The goal of the mechanistic approach is to predict the outcome before doing the experiment through a fundamental understanding of the dynamics of dissolution, ADME.

It is not suggested that the mechanistic approach will eliminate the need to do empirical experiments or eliminate the need to validate predicted outcomes through experimentation. However, as the science progresses, it is certainly a goal of the industry to predict outcomes to increase its success rate by eliminating ill conceived clinical studies, and a fundamental understanding of the ADME processes hold promise to this end.

Predicting dissolution falls under the realm of the formulation scientist, whereas methods to predict drug metabolism, toxicity, and efficacy generally do not. However, incorporating key aspects from all disciplines into the decision of what makes a successful drug product is likely to increase the quality of drug candidates. Here again, a mechanistically based approach holds the promise of wider applicability across diverse chemical structures and therapeutic areas. Mathematical models help bring the important parameters from each discipline together in a way so that more rational decisions can be made. As stated before, solubility and permeability are key parameters for the physical scientist working on dosage form development. Scientists involved in drug metabolism typically contribute estimates of drug clearance rates and volumes of distribution. Combining these two disciplines allows the prediction of drug plasma concentrations and whether or not the dose–exposure relationship will be linear or not. Although it is beyond the scope of this chapter, toxicologists and biological and clinical scientists can then review the predictions to see if projected plasma concentrations meet the needs for toxicological and clinical evaluation. This would ideally occur in project team meetings with representatives present from all disciplines.

The MAD analysis is mathematically simple, which is part of its appeal. However, more sophisticated models involve differential equations that do not necessarily have analytical solutions and, therefore, need to be solved numerically. The mathematical model to be presented as follows has the ability to simulate the kinetics of a polydisperse crystalline powder. This has wide applicability because of the prevalence of immediate-release dosage forms containing drug as a crystalline powder.

One of the assumptions of the model is that the crystalline drug particles are completely wetted and dispersed initially. The model does not describe the kinetics of wetting. However, by comparing theoretical simulations of the dissolution rate with actual dissolution from the dosage form, one can gain insight into the extent that wetting is slowing the rate of dissolution.

Validating and refining the model requires powder dissolution data that is independent of the effects of dispersion and wetting since this is an assumption of the model. This may require developing an experimental technique that uses a surfactant at a concentration that will not enhance solubility but will improve wetting. The technique may also require a brief period of vigorous mixing to achieve dispersion and wetting. High-quality data is required to validate dissolution theory as well as gain insight into some of the more elusive aspects such as how to handle hydrodynamics.

One of the goals of this chapter is to convince the reader that dissolution can be explained and predicted based on theory and that this is worthwhile in terms of shortening the time it takes to develop drug products. Perhaps the most dramatic way would be to show that, based on the solubility and permeability of a drug candidate, inherent absorption would never be good enough to allow the drug to become a product. Knowing this, project teams could decide whether to drop drug candidates and pursue others, or to commit resources in an attempt to overcome the solubility issue and accept the higher development cost and risk of failure in doing so. For the formulator, however, not knowing the effect of particle size on dissolution rate and absorption or whether poor disintegration or wetting is affecting the dissolution rate can lead to costly delays in development that could require the need to repeat toxicological and clinical studies.

Although the Biopharmaceutics Classification System (BCS) (13), discussed later, and MAD analysis are useful and attractive because of their simplicity, both are limited in terms of guidance that might be extracted from solubility, permeability, dissolution, and other pharmacokinetic data. Neither can describe the kinetics of absorption leading to insight into the effects of drug particle size and hydrodynamic conditions that would lead to a mechanistically based in vitro/in vivo correlation. They would also not allow one to make a rational estimation as to when dissolution samples should be taken and whether the dissolution test would be discriminating to significant differences in dosage forms. To do this, a more sophisticated model is needed such as the one described subsequently.

The dissolution rate of crystalline drug is proportional to its solubility, surface area, and diffusion coefficient. It is also dependent on the hydrodynamic conditions, but in a less well understood way. These relationships can be summarized in a Noyes–Whitney (14) type equation:

$$\frac{\mathrm{d}X_s}{\mathrm{d}t} = -\frac{DS}{h}\left(C_s - \frac{X_d}{V}\right) \tag{14}$$

where X_s is the mass of solid drug remaining at any time t, D is the drug diffusion coefficient, S is the drug surface area, h is the hydrodynamic diffusion layer thickness, C_s is the drug solubility, X_d is the mass of dissolved drug at any time t, and V is the volume of fluid in which the drug is dissolving.

In trying to solve the earlier equation, it should be noted that the drug surface area would not remain constant as drug dissolves after being released from typical dosage forms. The amount of dissolved drug would also not be constant. As a result, the rate of dissolution is continually changing. Dissolution testing is typically done under sink conditions; therefore, the X_d/V term is small compared to C_s so that former can be ignored. However, in trying to establish a mechanistically based in vitro/in vivo correlation, the assumption that sink conditions would exist in the GI tract is an especially bad one for poorly soluble drugs. Also, as will be shown, testing dissolution under sink conditions is not necessary and can make instrumental analysis of dissolution more difficult. What is required is a numerical solution of the Noyes–Whitney equation to make the application of the theory as general as possible by eliminating the need to make frequently bad assumptions in order to solve the equation analytically.

The following approach has been previously described (15,16). If one assumes that a drug particle has certain geometry, then surface area can be expressed in terms of drug mass if the drug density is known. The simplest geometry to use is spherical, although other geometries could be used (17). However, for the following derivations, spherical geometry will be assumed. The surface area at any given time can then be expressed by the following:

$$S = 4\pi r_t^2 N_0 \tag{15}$$

where r_t is the drug particle radius at any time t, and N_0 is the number of drug particles present initially. It will be shown later how one could handle a polydisperse drug powder, but for now, it will be assumed that all drug particles are exactly the same size and that they will all dissolve at the same rate. If this were the case, then the number of drug particles would not change with time until they completely dissolved at which time the number of particles would be zero.

The number of drug particles present initially can be calculated by dividing the initial mass of drug or dose by the mass of one drug particle:

$$N_0 = \frac{X_0}{\rho v_0} = \frac{X_0}{\rho \frac{4}{3}\pi r_0^3} \tag{16}$$

where ρ is the drug density, v_0 is the volume of one drug particle, X_0 is the initial mass of drug, and r_0 is the initial particle radius.

The previous equation can be solved for r_0 and then made dynamic by replacing X_0 and r_0 with their respective time-dependent variables X_S and r_t to yield:

$$r_t = \left(\frac{3X_s}{4\pi\rho N_0}\right)^{1/3} \tag{17}$$

By combining Equations 15, 16, and 17, surface area can be expressed as:

$$S = \frac{3X_0^{1/3}X_s^{2/3}}{\rho r_0} \tag{18}$$

Substituting the previous expression for surface area into the Noyes–Whitney equation gives:

$$\frac{dX_s}{dt} = -\frac{3DX_0^{1/3}X_s^{2/3}}{\rho r_0 h}\left(C_s - \frac{X_d}{V}\right) \tag{19}$$

If dissolution is occurring in a closed system, such as the dissolution vessel, then the amount of dissolved drug X_d is given by:

$$\frac{dX_d}{dt} = \frac{3DX_0^{1/3}X_s^{2/3}}{\rho r_0 h}\left(C_s - \frac{X_d}{V}\right) \tag{20}$$

Figure 1 shows the numerical calculation of X_S and X_d with time based on Equations 19 and 20, respectively. Because the simulation is for a closed system, the two curves representing X_S and X_d are symmetric. This would not be the case if one were to simulate drug dissolving in the GI tract while drug absorption was occurring.

Equations 19 and 20 are only able to handle a single particle size. To expand the application to polydisperse powders, it will be assumed that a polydisperse powder can be simulated as a collection of monodisperse powder

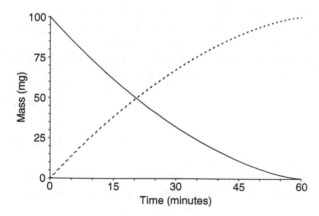

Figure 1 Simulation showing the dissolution of solid drug (*solid line*) from Equation 19 with the concomitant appearance of dissolved drug (*dashed line*) from Equation 20.

fractions indicated with a subscript i. Equations 19 and 20 become:

$$\frac{dX_{s_i}}{dt} = -\frac{3DX_{0_i}^{1/3}X_{s_i}^{2/3}}{\rho r_{0_i}h_i}\left(C_s - \frac{X_{d_T}}{V}\right) \tag{21}$$

$$\frac{dX_{d_i}}{dt} = +\frac{3DX_{0_i}^{1/3}X_{s_i}^{2/3}}{\rho r_{0_i}h_i}\left(C_s - \frac{X_{d_T}}{V}\right) \tag{22}$$

In solving these equations numerically using the Runge–Kutta method, the values of X_{S_i} and X_{d_i} are calculated at each step of the numerical method, the size of which can be selected as a trade-off between accuracy favored by smaller step sizes versus speed of calculation for larger step sizes. A typical step size would be approximately one second. After each step, the amount of solid and dissolved drug from each particle size fraction i would be totaled as follows:

$$X_{s_T} = \sum_{i=1}^{n} X_{s_i} \tag{23}$$

$$X_{d_T} = \sum_{i=1}^{n} X_{d_i} \tag{24}$$

where X_{s_T} and X_{d_T} are the total amount of solid and dissolved drug from all particle size fractions, respectively, and n is the number of particle size fractions. In Equations 23 and 24, it should be noted that all particles, regardless of their size, are dissolving based on the same concentration gradient:

$$\left(C_s - \frac{X_{d_T}}{V}\right) \tag{25}$$

as the value of X_{d_T} is updated after each step of the numerical calculation and the same value for X_{d_T} is used for each particle size fraction i.

In summary, simulation of the dissolution of a polydisperse powder is accomplished by treating it as a collection of monosized fractions. At time zero, dissolution is the fastest because there is the most surface area and the concentration gradient is the greatest. Using the Runge–Kutta numerical method and Equations 21 and 22, the amount of drug that has dissolved from each particle size fraction is calculated, and after each step of the simulation, Equations 23 and 24 are used to sum up all the contributions from each particle size fraction. The total amount of dissolved drug from all fractions is then used during the next step of the numerical method so that each particle size fraction is dissolving against the same concentration gradient. Dissolution slows with time because the surface area and concentration gradient are getting smaller.

Typically, milled drug powders are distributed lognormally by mass about some geometric mean particle size. This means that one can find a collection of particles of similar size that are smaller than the mean particle size and another collection of particles of similar size that are larger than the mean particle size, both collections of which are roughly equivalent in mass. However, since both collections are approximately equal in mass, the collection of smaller particles is made up of more particles and represents more surface area than the larger collection. As a result, the collection of smaller particles will dissolve faster and completely dissolve before the larger collection. Both the number and particle size distribution will change during the dissolution of a polydisperse powder, whereas only the particle size would change within each monosized fraction until complete dissolution was reached. At that point, the number within the fraction would become zero.

It is uncertain at what particle size one would be able to say that a particle is no longer a solid and that complete dissolution had occurred. However, particles calculated in the size range of molecular dimensions could probably be considered as completely dissolved. In computer simulation, without some statement as to when solid particles are completely dissolved, the calculated particle size will continue to decrease until the lower numerical limit of the computer system is reached. In agreement with the model described earlier for a polydisperse powder, it has been shown that during dissolution, the number of particles in a smaller particle size fraction decreased more rapidly relative to larger particle size fractions (18).

Table 3 shows how dissolution occurs as discussed earlier. For simplicity, the example of a polydisperse powder in Table 3 is made up of only three monodisperse fractions. In practice, more fractions would be needed to describe a more typical milled polydisperse drug powder. The simulation was done for 100 mg of drug with a solubility of 0.1 mg/mL dissolving in 1000 mL of water. With these parameters, the concentration of drug would be at the solubility when complete dissolution is reached. As can be seen in Table 3, the 100 mg of powder has an initial geometric mean of 25 μm containing most of the mass with smaller but equal amounts of mass at 6.25 and 100 μm. However, the 6.25 μm particle size fraction has the greatest number of particles and the most surface area per unit weight. In less than five minutes, the 6.25 μm particle size fraction has completely dissolved. The 25 μm particle size fraction took slightly more than two hours to dissolve, with the size, mass, and surface area decreasing proportionately as determined by geometry and density. Only the number of particles remained constant until dissolution was complete. The largest particle size fraction starting at 100 μm dissolved the slowest because it had the smallest surface area and also because the two smaller particle size fractions have dissolved more quickly, thereby reducing the concentration gradient environment for the remaining large particles. Even after 24 hours, the largest particle size fraction did not completely dissolve.

Evidence that dissolution occurs as described earlier can also be seen in the shape of actual dissolution data from a polydisperse powder. Figure 2 shows the powder dissolution of hydrocortisone (17). Experimental measurement of the original powder showed it had a geometric mean particle size of approximately 36 microns with a geometric standard deviation of 2.4. Two simulations based on the Noyes–Whitney theory are also shown. For one simulation, the powder

Table 3 Change in Various Drug Particle Parameters During Dissolution

Time (min)	Fraction 1	Fraction 2	Fraction 3
Particle size of each fraction (μm)			
0	6.25	25.0	100
5	0.00	23.4	99.4
30	0.00	17.5	97.3
60	0.00	12.5	96.0
120	0.00	3.66	94.8
1440	0.00	0.00	78.6
Drug mass in each fraction (mg)			
0	10.7	78.7	10.7
5	0.00	64.6	10.4
30	0.00	26.8	9.81
60	0.00	9.92	9.43
120	0.00	0.247	9.06
1440	0.00	0.000	5.17
Number of particles in each fraction/10,000			
0	6410	740	1.56
5	0	740	1.56
30	0	740	1.56
60	0	740	1.56
120	0	740	1.56
1440	0	0	1.56
Drug surface area in each fraction (mm^2)			
0	7.87	14.5	0.492
5	0.00	12.7	0.485
30	0.00	7.08	0.465
60	0.00	3.65	0.453
120	0.00	0.311	0.441
1440	0.00	0.00	0.304

Note: Simulation represents 100 mg of drug with a solubility of 0.1 mg/mL dissolving in 1000 mL of water.

was treated as a polydisperse powder using 16 monosized fractions to describe it. The mass and size of drug particles in each fraction were calculated based on the experimental data and the log-normal distribution function. For the other simulation, the powder was treated as a monodisperse powder with a size equivalent to the measured mean of 36 microns. The polydisperse simulation fitted the data much better than the monodisperse simulation as determined by the sum of residuals squared. Compared to the monodisperse simulation, the actual powder dissolved more quickly initially due to the presence of smaller particles with greater surface area, and slower later on, due to the presence of larger particles with less surface area. These phenomena, faster initial dissolution rate and slower final, are

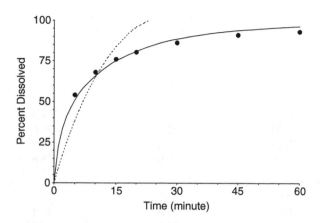

Figure 2   Dissolution of hydrocortisone (*solid circles*) and simulated dissolution with the drug modeled as a polydisperse powder (*solid line*) versus a monodispersed powder (*dashed line*). *Source*: From Ref. 17.

simulated better by modeling the drug as a polydisperse powder. Excellent agreement has also been reported between observed and simulated dissolution data for cilostazol at each of three median particle diameters of 13, 2.4, and 0.22 μm when modeled as polydisperse particles versus monodisperse (19).

Under certain special conditions, the described treatment of polydisperse powder dissolution would indicate that the mean particle size could increase; not because any particles were increasing in size, but because the smaller particles dissolve first, skewing the particle size distribution toward larger particles. As can be seen in Table 3, the initial geometric mean particle size was 25 microns. However, at 24 hours, all particles in fractions 1 and 2 have completely dissolved, leaving only particles in fraction 3. At that time, the particles in fraction 3 have gone from an initial value of 100 to 78.6 microns, leaving a mean particle size of 78.6 microns that is greater than the initial geometric mean of 25 microns.

One of the applications of trying to predict dissolution based on the Noyes–Whitney theory, solubility, and drug particle size is to identify potential formulation problems, such as wetting and slow disintegration. This is accomplished by comparing the predicted dissolution profile with the actual profile of the formulation. If the actual dissolution profile of the dosage form is similar to that predicted by theory, one could reasonably conclude that the formulation was disintegrating rapidly and that the surface area of the released drug particles were well wetted. However, dissolution slower than predicted should be investigated to determine the cause. To this end, drug powder dissolution in the absence of excipients but with the judicial use of surfactants and agitation to promote wetting but not to increase solubility or reduce particle size can help establish problems with agglomeration and poor wetting. Dissolution profiles faster than expected might indicate a change in drug form, resulting in a higher solubility or an increase in drug surface area, either of which might occur due to formulation processing.

The approach of using the dissolution theory described earlier to evaluate the dispersion process for furosemide has been reported (20). Good agreement between theoretical and experimental dissolution profiles were found when furosemide powders were dispersed by ultrasonication in a surfactant solution for all but the smallest of three batches of powder. The mean particle sizes for the batches were 3, 10, and 19 μm when particle size was measured after sonication. Without sonication, the particle sizes were measured to be 108, 38, and 27 μm corresponding to the post-sonicated measurements of 3, 10, and 19 μm, respectively. The relative order of the dissolution rates were also reversed before and after dispersion, indicating that comparing theoretical profiles with actual profiles would reveal the problem with agglomeration. For the smallest particle size batch of furosemide that did not agree well with the theoretically calculated dissolution rate, the drug particles were observed to agglomerate during dissolution, which would explain why the actual dissolution rate was slower than predicted by theory (M.M. De Villiers, personal communication, 2005).

Disintegration, wetting, and agglomeration should be understood and addressed by the formulator. If not, more variability in the in vitro/in vivo correlation is likely to result if a patient were to ingest something that might increase the wetting of a drug product that does not provide a surfactant itself. This would be analogous to adding a surfactant to the dissolution media instead of the formulation to achieve a desired dissolution profile. Again, theory can help the formulator identify potential dissolution problems.

The ability of the theory presented herein to simulate a polydisperse powder under nonsink conditions, which has been shown in studies that carefully address wetting and dispersion, challenges the conventional wisdom of conducting dissolution under sink conditions. The following example will be based on the physical properties of digoxin, whose bioavailability has been shown clinically to be dependent on its particle size (21). This dependency requires that drug particle size be controlled so that dissolution and bioavailability is consistent from batch to batch of drug product.

The question is whether to test dissolution under sink or nonsink conditions. Hypothetically, let it be assumed that the drug particle size specification calls for the drug powder to have a geometric mean particle size of 10 μm and a geometric standard deviation of 2. Figure 3 compares the simulated dissolution profiles of a 1 mg dose of drug that has a solubility of 0.05 mg/mL, similar in dose and solubility to digoxin. Profiles compare the simulated dissolution of a 1 mg dose in 900 or 90 mL of water for drug powders with geometric mean particle sizes of 10 and 20 μm, both with geometric standard deviations of 2. In Figure 3, dissolution is expressed as mass dissolved as a function of time with total dissolution occurring at the dose of 1 mg. The higher and lower solid-line profiles represent the dissolution of 10 and 20 μm powders, respectively, dissolving in 900 mL. The higher and lower dash-line profiles represent the dissolution of 10 and 20 μm powders, respectively, dissolving in 90 mL.

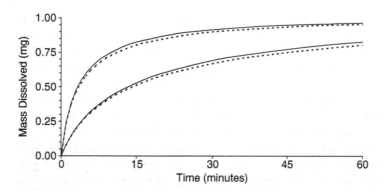

Figure 3 Simulated dissolution of a 1 mg dose with a solubility of 0.05 mg/mL in 900 mL (*solid lines*) versus 90 mL (*dashed lines*) for a drug with a mean particle size of 10 μ (*top two curves*) versus 20 μ (*bottom two curves*).

As expected, dissolution occurs faster in 900 versus 90 mL for both the 10 and 20 μm powders. However, the difference is not large and the ability of the simulated dissolution to differentiate the 10 versus 20 μm powder does not appear to depend on the volume of the water used in the dissolution test. In Figure 4, the same profiles are expressed as concentration instead of mass. Again, the dash-line profiles in Figure 4 are the dissolution profiles of drug dissolving in 90 mL of water, with the higher profile being the 10 μm powder and the lower being the 20 μm powder. The two solid-line profiles in Figure 4 are the dissolution profiles of drug dissolving in 900 mL of water, with the higher profile being the 10 μm powder and the lower being the 20 μm powder. Figure 4 shows

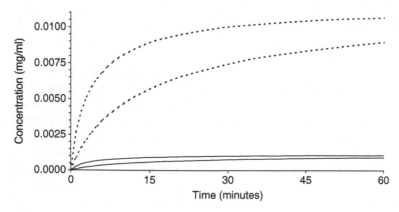

Figure 4 Simulated dissolution as in Figure 3 except only expressed as concentration instead of mass for a 1 mg dose with a solubility of 0.05 mg/mL in 900 mL (*solid lines*) versus 90 Ml (*dashed lines*) for drug with a mean particle size of 10 μ (*first and third curves from the top*) versus 20 μ (*second and fourth curves from the top*).

that the instrumental analysis would be much easier if dissolution were done in 90 mL of water because the resulting concentrations are 10 times higher. The conclusion of this example is that doing the dissolution testing under nonsink conditions in 90 mL of water versus sink conditions in 900 mL does not affect the ability to differentiate the particle size-dependent dissolution and makes the instrumental analysis of the difference easier.

Another advantage of adopting a modeling approach of simulation using a system of numerically solved differential equations is the ability to expand the model. Up to this point, the discussion has focused on describing events, such as dissolution and absorption, that occur on one side of the GI membrane with the fate of absorbed drug left undefined. However, by expanding the system of differential equations to describe the process of metabolism, tissue distribution, clearance, and excretion, the blood plasma versus time profile can be simulated in a dynamic way. This allows the coupling of dissolution and pharmacokinetics, with the absorption rate constant or permeability as the link, leading to an in vitro/in vivo correlation.

Figure 5 and the following equations will be used to illustrate how dissolution and pharmacokinetics can be combined in a dynamic way to provide a mechanistically based in vitro/in vivo correlation. The top cylinder in Figure 5 is meant to represent the GI tract. Inside the imaginary GI tract is shown a cylindrical plug on the left that is intended to represent a plug of GI fluid being propelled down the tract by peristalsis. The plug on the right is intended to

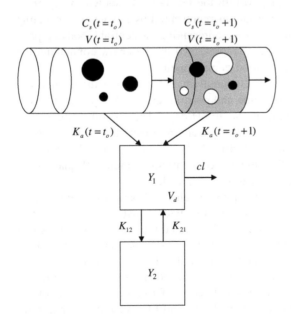

Figure 5   A schematic representation of the mathematical model described by Equations 26 to 33.

represent the same plug on the left only at a later time and new position due to peristaltic contractions causing the plug to slide down the tract. Inside the plug are shown various size circles that are meant to represent various size drug particles suspended and dissolving in the plug of GI fluid.

Several things happen as the plug of GI fluid moves down the GI tract. Initially, on the left, the plug is shown clear to represent that drug has not dissolved significantly. However, at a later time, as represented by the plug on the right, it should be noted that the black circles become smaller due to dissolution and one has dissolved completely. This would lead to an increase in the amount of dissolved drug in the plug, as represented by a darker shading of the plug on the right. If the dosage form did not release all of the drug particles at once, more drug particles would appear in the plug at a later time, as represented by the white circles. However, particles released at a later time may not dissolve at the same rate as previously released particles, because the concentration gradient may have changed from previously released dissolving drug.

Another change that could occur with time would be water absorption or secretion. In Figure 5, this change can be seen in the smaller size of the plug on the right relative to the plug on the left. In this example, water absorption has occurred to decrease the volume of GI fluid in the GI tract. Depending on the relative rate of drug versus water absorption, the concentration of dissolved drug in the plug of GI fluid could change and, therefore, affect the rate of dissolution.

At the same time, the absorption rate constant or permeability could be changing with position of the plug within the GI tract. As shown in Figure 5, the value of the absorption rate constant as represented by the arrow connecting the plug on the left with the square below, representing the central blood compartment, could change to a new value as represented by the arrow connecting the plug on the right with the central blood compartment.

Finally, although not visually apparent in Figure 5, the drug solubility within the plug could change with time and position due to changes in the composition of the GI fluid, most notably due to pH-dependent solubility changes in acidic and basic drugs passing from the low pH environment of the stomach to the higher pH environment of the small intestine. If the solubility falls below the concentration of drug dissolved in the plug, the concentration gradient (Equation 25) becomes negative and precipitation can be simulated.

How one would handle this interplay of rates and parameters and couple them with pharmacokinetics is shown in the later equations. As described before, the subscript i is used to index the individual particle size fractions that make up the entire distribution. Figure 5 only shows one particle in each of three different particles sizes with the understanding that, in reality, more fractions containing realistic number of particles can be handled mathematically. The subscript j is used to track when a collection of i particle size fractions is released from the dosage form. These fractions could be released over a fairly short time period to simulate disintegration or over a prolonged period of time to simulate a controlled-release formulation.

By definition, X_s and X_d are time-dependent variables. However, compared to the earlier presented dissolution equations, C_s, V, and the amount of drug released from the dosage form at any given time, X_{0ij}, can all be considered as time-dependent variables as indicated by the symbol for time t shown in parenthesis immediately following the parameter. It should be noted that the Runge–Kutta numerical method that would typically be used to solve the system of differential equations is an iterative method, providing the opportunity to change the values of time-dependent parameters at every step of the process, typically in the neighborhood of every second. Any time-dependent mathematical function could be used to determine the rate of change.

Absorption is simulated by adding the product of dissolved drug and a time-dependent absorption rate constant in Equation 27 and including a similar term in Equation 32 representing absorption into the central blood compartment. It should be noted that the amount of absorbed drug is calculated separately in Equation 28 from the calculation of drug actually reaching the central blood compartment in Equation 32. This is accomplished by adding the bioavailability term F in Equation 32 but not in Equations 27 or 28, reflecting the reality that some drugs can be absorbed to a greater percent, that is, from the GI tract, than would appear in the central blood compartment due to presystemic metabolism by the liver or even within the GI enterocytes.

$$\frac{dX_{s_{ij}}}{dt} = -\frac{3D\left(X_{0_{ij}}(t)\right)^{1/3} X_{s_{ij}}^{2/3}}{\rho h_i r_{0_i}}\left(C_s(t) - \frac{X_{d_T}}{V(t)}\right) \tag{26}$$

$$\frac{dX_{d_{ij}}}{dt} = +\frac{3D\left(X_{0_{ij}}(t)\right)^{1/3} X_{s_{ij}}^{2/3}}{\rho h_i r_{0_i}}\left(C_s(t) - \frac{X_{d_T}}{V(t)}\right) - K_a(t)X_{d_{ij}} \tag{27}$$

$$\frac{dX_{a_{ij}}}{dt} = K_a(t)X_{d_{ij}} \tag{28}$$

$$X_{s_T} = \sum_i^n \sum_j^t X_{s_{ij}} \tag{29}$$

$$X_{d_T} = \sum_i^n \sum_j^t X_{d_{ij}} \tag{30}$$

$$X_{a_T} = \sum_i^n \sum_j^t X_{a_{ij}} \tag{31}$$

$$\frac{dY_1}{dt} = K_a(t)FX_{d_T} - \left(\frac{cl}{V_d} + K_{12}\right)Y_1 + K_{21}Y_2 \tag{32}$$

$$\frac{dY_2}{dt} = K_{12}Y_1 - K_{21}Y_2 \tag{33}$$

The summation Equations 29, 30, and 31 are needed to calculate the total amount of solid, dissolved, and absorbed drug, X_{s_T}, X_{d_T}, and X_{a_T}, respectively, from all particle size fractions and all release times. It should be noted that all particles, no matter which fraction they came from or at which time they were released, dissolve against the same concentration gradient as determined by the difference between solubility and total concentration in solution X_{d_T}/V. This is necessary to provide a mechanistically realistic simulation of dissolution in vitro and in vivo.

The aforementioned parameters not previously defined include the mass of drug in the central blood compartment Y_1, the mass of drug in the peripheral or tissue compartment Y_1, clearance from the central compartment cl, and volume of distribution of the central compartment V_d.

Part of the value in using mathematical models to simulate dissolution, absorption, and pharmacokinetics comes from the training and insight that can be gained quickly and inexpensively relative to developing actual dosage forms and testing them in the clinic. Just as a pilot would use a flight simulator to learn how to fly, the formulation scientist can develop a feeling for how changes in various parameters affect the drug product without the large expense of laboratory and clinical studies. Writing differential equations also tests and strengthens understanding. Applying numerical methods to solve the equations eliminates the need to solve them analytically and frees the scientist to use equations that have no analytical solutions.

As with any good scientific approach, theory is validated through experimentation and modified to agree with confirmed experimental findings. An example of this would be the treatment of the hydrodynamics for simulating the dissolution of a polydisperse powder. Any dissolution model would need to explain why the rate of dissolution increases when the energy of stirring is increased. Including a diffusion layer thickness into the denominator of the Noyes–Whitney equation that becomes smaller when stirring is increased is used to explain this. As an initial attempt to simulate actual powder dissolution data (16,17), the assumption that the diffusion layer thickness is approximately equal to the radius of the dissolving particles was used (22). However, this assumption did not result in a good fit of the experimental data. Further search of the literature indicated that the diffusion layer thickness might plateau with increasing particle size (23). Applying this approach resulted in a much better fit of the data, and subsequent studies have confirmed the existence of a plateau diffusion layer thickness under typical drug dissolution testing conditions and particle sizes (24,25). However, the author anticipates that future work will lead to better understanding of the hydrodynamics of dissolving powders. More understanding is particularly needed in the in vivo environment.

The model described by Equations 26 to 33 has been applied to nifedipine to demonstrate how its physicochemical and pharmacokinetic properties could have been utilized to develop the controlled-release dosage form (26). The effect of particle size on dissolution and absorption of drug released from an osmotic pump controlled-release dosage form was simulated.

BIOPHARMACEUTICS CLASSIFICATION SYSTEM

The BCS (13) was introduced as a method to identify situations that might allow in vitro dissolution testing to be used to ensure bioequivalence in the absence of actual clinical bioequivalence studies. On the basis of the theoretical approach taken, solubility and intestinal permeability were identified as the primary drug characteristics that control absorption. This lead to classification of drugs into four broad groups as follows.

- Case 1: High solubility—high permeability.
- Case 2: Low solubility—high permeability.
- Case 3: High solubility—low permeability.
- Case 4: Low solubility—low permeability.

Neither the theoretical basis for the BCS nor the theoretical approach to model dissolution and absorption presented in this chapter have inherent boundaries that would naturally place any particular drug in one of the four BCS classes. However, both approaches do have regions of greater and lesser sensitivity to dissolution that warrant consideration as to whether in vitro dissolution could be used as a surrogate for bioequivalence testing. As pointed out in the theoretical justification for the BCS, the in vivo environment in which dissolution and absorption takes place has a high degree of variability. Out of necessity, boundaries for the BCS classes would have to error on the conservative side due to the uncertainties involved in estimating solubility and permeability in the GI tract.

The theoretical justification for the BCS did not clearly indicate where the boundaries between the four classifications should be. From a regulatory perspective, the boundaries are more clearly defined by the U.S. Food and Drug Administration, Center for Drug Evaluation and Research (27). This information is entitled "The Biopharmaceutics Classification System (BCS) Guidance" and provides guidance for "Waiver of In-vivo Bioavailability and Bioequivalence Studies for Immediate Release Solid Oral Dosage Forms Based on a Biopharmaceutics Classification System." The guidance describes the requirements for a drug to be considered highly soluble, highly permeable, and rapidly dissolving. It also offers a variety of methods for establishing that a drug is highly soluble or permeable. Further restrictions are placed on a request for a waiver of bioequivalence testing that include the requirement that the drug have a wide therapeutic window and that excipients used in the dosage form must have been used in a previously approved immediate-release solid dosage form by the Food and Drug Administration.

Although the names of the four BCS classes do not indicate so, dose is an essential piece of information used in the calculation to determine whether a drug can be considered as highly soluble as described by the BCS guidance. Its importance follows from the theoretical basis for the BCS and the dissolution theory presented in this chapter, as drug surface area in the Noyes–Whitney theory is dose-dependent. For a drug to be considered highly soluble, the highest dose must be soluble in 250 mL of water or less over a pH range of 1 to 7.5. The significance of dose has been pointed out by comparing digoxin and griseofulvin as drugs that have roughly similar physical properties of solubility and permeability, but vary considerably with respect to dose (15). As a result, the high dose of digoxin would dissolve in 250 mL of water, whereas the high dose of griseofulvin would not. Therefore, according to the BCS guidance, digoxin would be considered as highly soluble and griseofulvin would not. It should be noted that the BCS-based biowaiver does not apply to narrow therapeutic range drugs like digoxin (28).

The BCS was developed on the theory that drug dissolution is controlled by solubility and drug surface area as defined by dose and drug particle size. In accepting the BCS, it follows that there should be a theoretical rate of drug dissolution given the solubility, dose, particle size, dissolution volume, and hydrodynamic conditions. This conclusion is also the intent of the modeling of a polydisperse drug powder presented herein. The BCS guidance is also intended to apply only to immediate-release solid dosage forms. As such, the dosage form should disintegrate within a few minutes when exposed to water to release drug particles.

Dissolution theory allows the formulator to calculate the rate of drug dissolution and compare it to actual experimental dissolution data. Discrepancies can then be investigated, which could be due to the effects of disintegration, wetting, inaccurate particle size information, or faulty theory. Dissolution of well dispersed, wetted drug particles in the absence of the formulation can also be done for comparison with the dissolution data from the solid dosage form and checked against the theoretical dissolution rate. This ensures that the formulator understands how the dosage form is behaving.

Given the assumption that the purpose of an immediate-release dosage form is to rapidly disintegrate to release well-dispersed and wetted drug particles, establishing the desirable drug particle size distribution remains as an important task under the control of the formulator. The question that needs to be addressed is what should the drug particle size be to rapidly dissolve according to the BCS Guidance? To explore this question, two hypothetical drugs can be compared. Both are high permeability drugs with absorption rate constants of 0.03 reciprocal minutes. One has a dose of 250 mg with a solubility of 1 mg/mL, and the other has a dose of 2.5 mg with a solubility of 0.01 mg/mL. As such, both drugs will just dissolve in 250 mL of water and are on the boundary of being considered Case 1 drugs: high solubility, high permeability.

Figure 6 compares the simulated percent of dose absorbed for both drugs, each simulated with a geometric mean particle size of 5 and 25 microns. A mean of 5 micron would be typical of drug that had been jet-milled, whereas 25 microns would not be an unusual particle size for drug milled by other conventional mills used in the pharmaceutical industry. The top two curves, representing the 250 mg dose at a solubility of 1 mg/mL, show little difference in the absorption profile

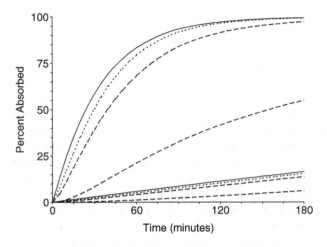

Figure 6 Simulated absorption of drug. Legend for Figure 6 is shown in Table 4.

for particle sizes of 5 and 25 microns. However, in the third and fourth curves from the top, the simulated absorption profile for the 25 micron particle size representing a 2.5 mg dose with a solubility of 0.01 mg/mL (lowest curve) is very different than the 5 micron particle size for the same dose and solubility. The conclusion drawn from this theoretical set of simulations is that drugs in the same high solubility, high permeability BCS class do not have the same sensitivity to drug particle size with regard to dissolution. It should be noted that the 2.5 mg dose, 0.01 mg/mL solubility drug that was simulated to be sensitive to particle size has similar properties to digoxin whose absorption has been shown to be sensitive to drug particle size.

As mentioned earlier, the BCS Guidance requires that the drug product be rapidly dissolving. More specifically, 85% or more of the drug substance must

Table 4 Parameters for Simulations in Figure 6

Dose (mg)	Solubility (mg/mL)	Absorption rate constant (1/min)	Drug particle size (μm)	Line style
250	1	0.03	5	solid
250	1	0.03	25	dot
2.5	0.01	0.03	5	dash
2.5	0.01	0.03	25	dot-dash
250	1	0.001	5	solid
250	1	0.001	25	dot
2.5	0.01	0.001	5	dash
2.5	0.01	0.001	25	dot-dash

Note: Simulations in Figure 6 correspond to the rows in Table 4 in the same order from top to bottom, respectively.

dissolve in 30 minutes using USP apparatus I or II in a volume of 900 mL or less. When the simulations mentioned earlier were repeated in 900 mL with no absorption, both the 5 and 25 micron 250 mg dose with a solubility of 1 mg/mL would meet the 85% rapidly dissolving criterion as well as the 5 micron 2.5 mg dose with a solubility of 0.01 mg/mL. For the 25 micron 2.5 mg dose, 0.01 mg/mL solubility, only 23% of the dose was simulated to be dissolved at the 30 minute time point. The practical conclusion would be that the 2.5 mg dose, 0.01 mg/mL solubility drug would have to be milled to approximately 5 micron to qualify for the biowaiver according to the BCS guidance. Therefore, the rapidly dissolving requirement of the BCS Guidance provides another safety check by forcing a tighter particle size specification for drugs that are more sensitive to the effect of particle size on dissolution.

The same simulations as mentioned earlier can be repeated using an absorption rate constant of 0.001 instead of 0.03 reciprocal minutes, changing the drugs from Case 1 to Case 3: high solubility—low permeability, to give the lower set of four curves shown in Figure 6. The absolute differences between the Case 3 simulations are smaller than those for the Case 1 simulations. This brings up the question as to why Case 3 drugs are not eligible for a biowaiver with a single point dissolution specification of 85% at 30 minutes. If Case 3 dosage forms are rapidly dissolving, it is unlikely that variability in absorption is due to formulation effects. This point has also been made in the original theoretical justification for the BCS. However, only Case 1 drugs are currently eligible for a biowaiver based on the BCS Guidance.

CONCLUSION

Modeling approaches to simulate dissolution, ADME provide tools to help the pharmaceutical scientist understand these processes and to guide decisions on drug selection and development. Application of dissolution and absorption theory has led to the BCS, which holds promise in reducing the burden of demonstrating bioequivalence by using a single-point in vitro dissolution test as a surrogate for in vivo clinical studies for Case 1 drugs. Although the regulatory benefits are limited to Case 1 drugs, application of modeling tools in the pharmaceutical industry may reduce the time and expense of developing new drugs of all classes at each step of the discovery to market process. The theory highlights the importance of solubility, permeability, and pharmacokinetics, and brings these elements together in a way that allows for comprehensive decisions, from avoiding drugs that are likely to be difficult to develop, to setting drug particle size specification to ensure consistent dissolution, to pursuing solubility enhancing or controlled-release formulation.

REFERENCES

1. Hidalgo IJ, Raub TJ, Borchardt RT. Characterisation of the human colon carcinoma cell line (Caco-2) as a model system for intestinal epithelial permeability. Gastroenterology 1989; 96(3):736–749.

2. Kansy M, Senner F, Gubernator K. Physicochemical high throughput screening: parallel artificial membrane permeability assay in the description of passive absorption processes. J Med Chem 1998; 41:1070–1110.
3. http://www.intellipharm.com/intellipharmka.htm (accessed May 2005).
4. Fagerholm U, Johansson M, Lennernäs H. Comparison between permeability coefficients in rat and human jejunum. Pharm Res 1996; 13(9):1336–1342.
5. Lennernäs H, Knutson L, Knutson T, et al. Human effective permeability data for atenolol, metoprolol, propranolol, desipramine and carbamazepine to be used in the proposed biopharmaceutical classification for IR-products. Eur J Pharm Sci 1996; 4(1001):S69.
6. Takamatsu N, Kim O-N, Welage LS, et al. Human jejunal permeability of two polar drugs: cimetidine and ranitidine. Pharm Res 2001; 18(6):742–744.
7. Lennernäs H. Human intestinal permeability. J Pharm Sci 1998; 87(4):403–410.
8. Fujimura A, Kumagai Y, Sugimoto K, et al. Circadian influence on effect of propranolol on exercise-induced tachycardia in healthy subjects. Eur J Clin Pharmacol 1990; 38(2):133–137.
9. Lin JH, Yamazaki M. Clinical relevance of P-glycoprotein in drug therapy. Drug Metab Rev 2003; 35(4):417–454.
10. Chiou WL, Chung SM, Wu TC, et al. A comprehensive account on the role of efflux transporters in the gastrointestinal absorption of 13 commonly used substrate drugs in humans. Int J Clin Pharmacol Ther 2001; 39(3):93–101.
11. Johnson KC, Swindell AC. Guidance in the setting of drug particle size specifications to minimize variability in absorption. Pharm Res 1996; 13(12):1795–1798.
12. http://www.intellipharm.com/intellipharmmad.htm (accessed May 2005).
13. Amidon GL, Lennernäs H, Shah VP, et al. A theoretical basis for a biopharmaceutic drug classification: the correlation of in vitro drug product dissolution and in vivo bioavailability. Pharm Res 1995; 12(3):413–420.
14. Noyes AA, Whitney WR. The rate of solution of solid substances in their own solutions. J Am Chem Soc 1887; 19:930–934.
15. Dressman JB, Fleisher D. Mixing-tank model for predicting dissolution rate control of oral absorption. J Pharm Sci 1986; 75(2):109–116.
16. Hintz RJ, Johnson KC. The effect of particle size distribution on dissolution rate and oral absorption. Int J Pharm 1989; 51:9–17.
17. Lu ATK, Frisella ME, Johnson KC. Dissolution modeling: factors affecting the dissolution rates of polydisperse powders. Pharm Res 1993; 10(9):1308–1314.
18. Simões S, Pereira de Almeida L, Figueiredo M. Testing the applicability of classical diffusional models to polydisperse systems. Int J Pharm 1996; 139:169–176.
19. Jinno J, Kamada N, Miyake M, et al. Effect of particle size reduction on dissolution and oral absorption of a poorly water-soluble drug, cilostazol, in beagle dogs. J Control Release 2006; 111:56–64.
20. De Villiers MM. Influence of agglomeration of cohesive particles on the dissolution behaviour of furosemide power. Int J Pharm 1996; 136:175–179.
21. Jounela AJ, Pentikäinen PJ, Sothmann A. Effect of particle size on the bioavailability of digoxin. Eur J Clin Pharmacol 1975; 8:365–370.
22. Higuchi WI, Hiestand EN. Dissolution rates of finely divided drug powders. I. Effect of a distribution of particle sizes in a diffusion-controlled process. J Pharm Sci 1963; 52:67–71.
23. Harriott P. Mass transfer to particles. I. Suspended in agitated tanks. AIChE J 1962; 8:93–101.

24. Bisrat M, Anderberg EK, Barnett MI, et al. Physicochemical aspects of drug release. XV. Investigation of diffusional transport in dissolution of suspended, sparingly soluble drugs. Int J Pharm 1992; 80:191–201.
25. De Almeida LP, Simões S, Brito P, et al. Modeling dissolution of sparingly soluble multisized powders. J Pharm Sci 1997; 86(6):726–732.
26. Johnson KC. Dissolution and absorption modeling: model expansion to simulate the effects of precipitation, water absorption, longitudinally changing intestinal permeability, and controlled release on drug absorption. Drug Dev Ind Pharm 2003; 29(8):833–842.
27. http://www.fda.gov/cder/OPS/BCS_ guidance.htm (accessed July 2005).
28. http://www.fda.gov/cder/guidance/3618fnl.htm (accessed July 2005).
29. Avdeef A, Berger CM, Brownell C. pH-metric solubility 2: correlation between the acid-base titration and the saturation shake-flask solubility-pH methods. Pharm Res 2000; 17(1):85–89.
30. Ran Y, Zhao L, Xu Q, et al. Solubilization of cyclosporin A. AAPS Pharm Sci Tech 2001; 2(1) article 2 (http://www.aapspharmscitech.org/).
31. Hesselink DA, van Gelder T, van Schaik RHN, et al. Population pharmacokinetics of cyclosporine in kidney and heart transplant recipients and the influence of ethnicity and genetic polymorphisms in the MDR-1, CYP3A4, and CYP3A5 genes. Clin Pharmacol Ther 2004; 76(6):545–556.
32. Chiou WL, Kyle LE. Differential thermal, solubility, and aging studies on various sources of digoxin and digitoxin powder: biopharmaceutical implications. J Pharm Sci 1979; 68(10):1224–1229.
33. Parker RB, Yates CR, Soberman JE, et al. Effects of grapefruit juice on intestinal P-glycoprotein: evaluation using digoxin in humans. Pharmacotherapy 2003; 23(8):979–987.
34. Mosharraf M, Nyström C. Apparent solubility of drugs in partially crystalline systems. Drug Dev Ind Pharm 2003; 29(6):603–622.
35. Yazdanian M, Briggs K, Jankovsky C, et al. The "high solubility" definition of the current FDA guidance on biopharmaceutical classification system may be too strict for acidic drugs. Pharm Res 2004; 21(2):293–299.
36. Swanson DR, Barclay BL, Wong PSL, et al. Nifedipine gastrointestinal therapeutic system. Am J Med 1987; 83(suppl. 6B):3–9.

2

Pharmacokinetics: Basics of Drug Absorption from a Biopharmaceutical Perspective

Sandhya K. Apparaju and Srikanth C. Nallani
Food and Drug Administration, Silver Spring, Maryland, U.S.A.

INTRODUCTION

For a drug to exert the desired pharmacological effect, therapeutic concentrations of the drug need to be available at the cellular sites of action. In practice, systemic concentrations are measured and this systemic bioavailability is used as an indicator of whether adequate concentrations of the drug are achieved at the final site of action. Absorption from the site of administration is one of the foremost hurdles that a drug molecule encounters before it can reach systemic circulation. Depending on the route of drug administration, several anatomical barriers to drug absorption and resultant systemic bioavailability exist. The ability of a drug to cross these barriers depends not only on the physicochemical properties of the drug molecule and the formulation characteristics (drug product design), but also on various physiological processes at the site of absorption. In addition, interactions between drug–drug, drug–food, drug–disease, as well as the presence of various drug-metabolizing enzymes, uptake and efflux transport proteins at the membrane barriers, influence drug absorption. Knowledge of factors

This book chapter was written by Drs. Apparaju and Nallani in their private capacity. No official support or endorsement by the Food and Drug Administration is intended or should be inferred.

influencing drug absorption is critical for choosing an appropriate route of administration and development of an optimal dosage form.

In this chapter we will discuss the basics of drug absorption from a biopharmaceutical perspective, that is, those factors whether physiological or physicochemical that may influence the ability of a drug to get absorbed from the site of administration into the systemic circulation.

ROUTES OF DRUG ADMINISTRATION

Choosing an optimal route for drug administration is one of the most important decisions in drug development, as this choice may influence several end points that include:

- Drug stability
- Size and frequency of dosing
- Rate and extent of drug absorption
- Time for onset of therapeutic response
- Duration and magnitude of the therapeutic response
- Adverse event profile
- Patient compliance and convenience

Commonly employed routes of drug administration intended for systemic effect of the drug include:

- Parenteral (intravenous, intramuscular, subcutaneous, intra-arterial)
- Enteral (buccal, sublingual, oral, rectal)
- Percutaneous (or transdermal)
- Inhalation
- Intranasal

The parenteral routes of drug administration are associated with high bioavailability due to the absence of (intravenous) or limited (intramuscular) barriers to drug absorption. Bioavailability from extravascular routes is usually lower and also highly variable. This results from the presence of significant barriers to drug absorption and also a complex interplay of multiple drug and physiological factors.

Parenteral

Intravenous route is the fastest and most predictable route of drug administration as it avoids the rate-limiting absorption phase associated with other routes. The onset of the desired pharmacological effect is prompt and is, therefore, especially beneficial in case of emergencies. However, because of the rapid appearance of peak drug concentrations in systemic circulation, this route of drug administration can also be risky. Several drugs are administered by intravenous routes including analgesic, anesthetic, antibiotic, anticancer, and antiepileptic drugs. Intravenous infusions are particularly desirable when peak drug concentrations need to be avoided and the drug concentrations in blood need to be controlled within a desirable range to maximize therapeutic benefit and to minimize toxicity.

Typically drug absorption from the subcutaneous and intramuscular routes depends on the solubility of drug in surrounding fluids and the blood flow to the site of administration. Absorption is usually rapid compared to oral route or can be modified to achieve a sustained release when desired (e.g., intramuscular depot injection of Leuprolide).

Oral

Drug absorption from the gastrointestinal (GI) tract is determined by transport across both the GI epithelial membrane as well as the endothelial membrane of the capillary walls. Drug absorption across the epithelial barriers typically occurs via passive diffusion. For some drugs such as L-dopa and 5-fluorouracil, drug absorption across intestinal epithelial cells involves transport by endogenous carrier proteins. In general, drug absorption from the GI tract is influenced by several factors including:

- Physicochemical properties of the drug: particle size, lipophilicity, solubility, polymorphism, etc.,
- Physiological variables of the gastrointestinal tract: gastric emptying, small intestinal transit time, pH, presystemic elimination, drug efflux, etc., and
- Propensity for drug-, food-, or disease-related interactions.

Therefore, the net bioavailability of a drug as well as therapeutic effectiveness is highly variable compared to parenteral routes. However, due to high safety, patient convenience, compliance, and low cost, it is still by far the most preferred route.

Transdermal

This route of administration makes use of the readily accessible surface of human skin for achieving drug delivery into systemic circulation. The drug products typically are multi-layered patches that include either a layer of the active drug separated from the skin by a layer of rate-controlling polymer (membrane or reservoir-type) or where the active drug is dispersed in a matrix of rate-controlling material (matrix-type). This route of administration allows a noninvasive, zero-order drug input that delivers drug to the body at a constant rate with minimal fluctuations in the resultant serum drug concentrations. In addition, bioavailability via this route is improved compared to oral route, as it avoids the firstpass metabolism and drug instability in the GI environment. Patient compliance due to the noninvasive and less frequent nature of drug administration is another advantage. However, drug absorption may become erratic when the skin integrity is compromised due to an infection, injury, or in cases where the patch itself might become damaged. Contraceptives, hormone replacement therapies, analgesics, and smoking-cessation products are some of the commonly delivered treatments by means of this route. While most patches employ passive transdermal technology, i.e., drug delivery across a concentration gradient, due to the high dependence of this pathway on the physicochemical properties of the molecule including its molecular size and polarity, the number of molecules that are amenable to this route are limited. Use of active transdermal technologies that employ external driving forces to deliver a molecule through the skin, makes it feasible to achieve transport of large molecules such as peptides, proteins, and drugs with unfavorable physicochemical properties, which otherwise could not penetrate intact skin. Examples include the use of sonophoresis, iontophoresis, electroporation, heat or thermal energy.

Inhalation

This route is primarily used to treat diseases of the respiratory tract (e.g., bronchodilators and corticosteroids) and for the administration of anesthetics. It allows localized delivery of drugs for a faster onset of action and reduced systemic toxicity. Systemic absorption can also occur readily from the lungs due to the large surface area of the alveoli and can be employed for systemic drug therapy. It is also a noninvasive alternative for delivery of macromolecules such as peptides and proteins. Absorption from this route is also dependent upon the adaptability of a drug molecule to formulation processes critical for maximizing the inhalation and deposition. These include:

- Particle size reduction,
- Dispersability to form a fine aerosol,
- Compatibility with excipients including solvents, suspending agents, permeation enhancers, polymers to achieve sustained release, and
- Drug stability during formulation development and upon inhalation.

Intranasal

The intranasal route is commonly employed to achieve local therapeutic benefit for conditions such as allergic rhinitis, asthma (e.g., nasal decongestants, antihistamines, corticosteroids) or for local anesthetic effect (e.g., lidocaine). However, drugs administered into the nasal cavity can also be absorbed into the systemic circulation by means of the richly perfused and large surface area of the nasal mucosa. Absorption by this route bypasses the first pass metabolism, thereby allowing greater drug absorption to occur compared to the oral route for certain drugs. Because of the potential for improved bioavailability and a rapid onset of therapeutic effect, several drugs are administered via this route. These include selective 5-hydroxytryptamine receptor agonists for migraine relief, desmopressin for nocturia, insulin for diabetes, calcitonin for osteoporosis, and live influenza virus vaccine. Lipophilic drugs in general are readily absorbed through the nasal mucosa with bioavailability approaching close to that from an intravenous injection.

Absorption of polar drugs as well as large molecular weight peptides and proteins from nasal mucosa may be a challenge due to their low membrane permeability and also due to the rapid mucociliary clearance mechanism. Absorption enhancers such as surfactants, bile salts, phospholipids, cyclodextrins, chitosan, mucoadhesive agents have been demonstrated to improve the uptake of such therapeutic agents due to their varying effects on the nasal mucosa. Such effects include changes in the integrity of phospholipid bilayers, effects on the tight intercellular junctions of the epithelial membrane, inhibition of enzymatic activities, or simply by prolonging the residence time of the drug in the nasal cavity to promote absorption. In recent years, nasal route has also been explored as a pathway to achieve therapeutic drug concentrations in the brain, which may benefit treatment of central nervous system indications such as Parkinson's and Alzheimer's diseases. Some of the disadvantages of this route include the need for instilling small concentrated volumes of the drug, a smaller residence time, and irritation of the nasal mucosa.

Buccal and Sublingual

Drugs can be absorbed from the sublingual (floor of the mouth) and buccal (cheeks) mucosal membranes of the oral cavity because of the high permeability of these linings and their ability to retain a drug for prolonged duration. Drug absorption occurs directly from the oral cavity, bypassing the firstpass metabolism and can result in rapid achievement of therapeutic concentrations in the plasma and prompt therapeutic benefits. Nitroglycerin therapy for acute angina pectoris is a popular example of a drug that is benefited from this route of administration, resulting in pharmacological benefit after only one to two minutes. Other drugs that demonstrated benefit by means of this route include morphine, verapamil, nifedipine, captopril, and 17-beta estradiol.

Rectal

Rectal mucosa can be used for achieving systemic delivery of drugs such as antiemetics, antiepileptics, pain relievers, sedatives, and anesthetics. Although absorption from this route can be quite erratic, it is nevertheless a beneficial route for use in pediatric patients and for patients who are unable to swallow or retain an oral dosage form. The absorption from this route is influenced by several factors including:

- Formulation characteristics
- Drug concentration
- Site of drug delivery
- Temperature and pH of the rectal cavity
- Presence of stools in the rectum
- Retention of the dosage form and
- Differences in venous drainage of the rectum

DRUG TRANSPORT ACROSS BIOLOGICAL BARRIERS

Physiological barriers to drug absorption exist in the form of membrane structures that prevent the cellular uptake of most charged, hydrophilic, large lipophilic molecules, or molecules attached to protein structures. Drug absorption from the site of administration into the systemic circulation as well as drug absorption from the systemic circulation to the sites of action is regulated by the presence of these cellular membranes.

The Fluid-Mosaic model describes the cell membrane as a fluid, dynamic structure comprised of phospholipids, proteins, and carbohydrates (Fig. 1).

In this model the phospholipids molecules are hypothesized to be arranged in a bilayer, with the polar groups in each layer oriented away from the center forming the inner and outer surfaces of the membrane, and the hydrocarbon chains of each layer aligned in the middle to form the core hydrophobic domain. This central hydrophobic domain represents the primary barrier to absorption of water-soluble molecules including drugs across cellular structures. While small, lipophilic molecules are able to diffuse easily through this lipophilic cell membrane, the presence of pores within the membrane allows the passage of essential molecules including water into and out of the cell. In addition, presence of interspersed transmembrane protein structures, either partially or completely embedded within the bilayer, enables the transport of some essential hydrophilic molecules. The presence of peripheral protein structures provide for structural integrity of the membrane.

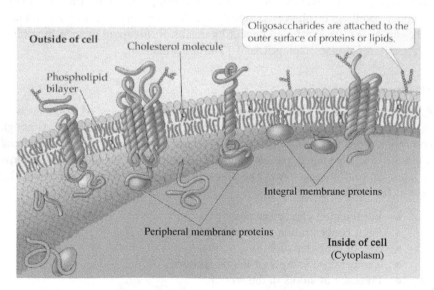

Figure 1 An illustration of the Fluid-Mosaic model. *Source*: Courtesy of Sinauer Associates, Inc., © 1998.

In general, passage of drugs across cell membranes may involve transcellular (movement across the cell) or paracellular (movement between the cells) processes. Paracellular process is passive (not requiring external energy) and depending only on the size and shape of the drug molecule. Transcellular uptake processes can be passive, active, or facilitated, and are responsible for the membrane permeability of most drugs with a molecular weight above 200 daltons.

Passive diffusion or simple diffusion is the most common mechanism of drug uptake across cell membranes. It is a nonenergy dependent, nonsaturable, concentration gradient-dependent spontaneous movement of drug molecules across cell membranes. As described by Fick's Law of Diffusion, which states that the rate of diffusion, or flux J of a species is proportional to the concentration gradient, movement of drug molecules occurs from a high concentration region to a low concentration region, and is influenced by various factors inherent to the drug molecule and the membrane such as the lipid solubility of the drug, the surface area, and thickness of the exposed barrier, as defined by the following equation:

$$J = -DAK/L(C_o - C_i)$$

where J is the rate of diffusion, i.e., rate of solute flux from C_o to C_i, which represents the concentration difference between the outside and inside of the membrane structure across which the diffusion occurs; D is the diffusion coefficient (i.e., speed with which a molecule can move across the membrane as influenced by its size and shape); A is the surface area of the membrane (larger the surface area, greater is the amount of drug that passes through); K is the lipid-water partition coefficient, i.e., a measure of lipid solubility, one of the major determinant of the pharmacokinetic characteristic of drugs (K increases with increasing lipid solubility; because the cell membranes are lipophilic in nature, lipid-soluble drugs diffuse more freely than relatively lipid-insoluble drugs); and L is the membrane thickness (as thickness increases, rate of diffusion decreases); the term DAK/L is collectively known as the permeability constant "P," which is the number of molecules crossing the membrane per unit area in unit time. As the value of the permeability constant increases, the rate of drug equilibration across membranes increases.

Facilitated diffusion is a nonenergy dependent (passive), saturable process that is mediated by carrier proteins or ion channels (voltage-gated or ligand-gated) that allow transport of selective molecules or ions down a concentration gradient. Because of the involvement of protein structures and hydrophilic channels this process is selective for a closely related group of molecules and is subject to competitive inhibition. Facilitated diffusion allows for the cellular passage of highly polar or ionized molecules such as glucose, sodium and chloride ions, and so on. The uptake of excess glucose from the blood and storage in the liver cells following a high carbohydrate meal as well as release of stored glycogen into the systemic circulation during hypoglycemic conditions involves facilitated diffusion by transporters across the hepatic cell membranes.

Vesicular transport involves the engulfing of fluid or small particulate (pinocytosis) and large particulate (phagocytosis) material by vesicles and vacuoles, which then fuse with the cell membranes to either transport/store (endocytosis) or release (exocytosis) molecules into and out of the cells. The release of insulin by pancreatic cells is a classic example of exocytotic vesicular transport.

Ion pairing occurs when highly charged ions, which are otherwise not permeable across cell membranes, form lipid-soluble complexes with oppositely charged ions. This results in the formation of a neutral molecule that has a polarity and hydrophobicity more suitable for partitioning into lipid phase of the membranes, than each ion by itself.

Active transport is an energy-dependent saturable process facilitating drug transport against concentration gradient. Transporters such as P-glycoprotein, organic anion transport protein, OATP, and System L transporters are known to affect absorption of endogenous and exogenous chemicals, including drugs from GI tract. P-glycoprotein (P-gp or MDR1) is a member of the efflux transporter classified as the ATP-binding cassette transporters. It is present in the brush border membrane of small intestine enterocytes and its efflux transport activity is responsible for low systemic absorption of a number of drugs. Its expression and efflux activity is also noted in liver, kidney, and endothelial cells of the blood-brain barrier and placenta. OATP subtype A, is expressed in the intestine and its uptake activity is responsible for absorption of drugs (1). Since P-gp and OATP are coexpressed in the intestine and share specificity for a variety of substrate drugs, coordinate activity of the efflux and uptake activities might determine the extent of drug absorption. System L consists of several basolateral membrane transporters, which mediate Na^+-independent transport of large neutral amino acids through the epithelial cells of blood-tissue barriers [blood-brain barrier (BBB) and placenta], small intestine, and renal proximal tubules. For example, gabapentin, a gamma amino butyric acid (GABA) analogue, is a substrate for this transporter and its absorption is limited at higher doses as explained by the saturation of this low capacity transporter activity.

FACTORS INFLUENCING DRUG ABSORPTION

Physicochemical Factors Affecting Drug Absorption

Drug absorption is primarily a function of drug dissolution and membrane permeability of the dissolved drug. For hydrophilic or polar drugs, membrane permeability usually is the rate-limiting step for drug absorption, while for those drugs (small lipophilic or uncharged) that can easily penetrate the membrane barriers, drug dissolution in the aqueous GI fluids may be rate limiting. Therefore, physicochemical formulation as well as physiological factors of the GI tract that alter drug dissolution and permeability will ultimately influence the dose absorbed from the administration site into the systemic circulation. The Noyes-Whitney equation, which is an extension of the Fick's First Law of

Diffusion, suggests that the rate of dissolution of a drug in the GI tract is driven by its solubility in the GI contents and the surface area of the drug exposed to lumenal fluids. Factors that can influence drug solubility in the GI fluids include the drug's polymorphism, lipophilicity, and pKa in relation to the pH profile of the GI tract.

Polymorphism

Polymorphism is the ability of a substance to exist in more than one crystalline form, each differing in their arrangements and/or conformations of the molecules within the crystal lattice. Due to differences in their energies, polymorphs usually differ in physico-chemical properties including solubility and therefore may result in varying dissolution rates. A very large number of pharmaceuticals are shown to exhibit polymorphism, with solubility being affected by the nature of the polymorph for several drugs including diflunisal, chloramphenicol palmitate, glibenclamide, ketorolac, ranitidine, and ritonavir. Characterization of these polymorphs and understanding their potential for interconversion during the manufacturing or storage of a pharmaceutical is important to produce a pharmaceutical product of consistent bioavailability and biological effectiveness.

pKa and pH

According to the pH-partition theory, uncharged or neutral molecules are able to pass biological membranes (i.e., undergo passive diffusion) more efficiently due to their greater lipid solubility compared to ionized molecules that are not lipid soluble. Because most drugs are either weak acids or bases, the rate of permeation of these drugs is largely governed by the relative concentrations of their ionized and unionized species at the cell membrane. This in turn is dependent on the pKa of the drug as well as the pH of the environment as described by the Henderson–Hasselbalch equations.

For a weak acid that dissociates in presence of water by donating a proton to form an ionized species, Henderson–Hasselbalch equation is defined as

$$pH = pKa + \log[\text{ionized/unionized}],$$

where Ka is the dissociation constant of the weak acid (lower the pKa, stronger is the acid). Therefore, for a weak acid as pH increases, the concentration of unionized species decreases.

For a weak base that acts as a proton acceptor to form its molecular (unionized) species, Henderson–Hasselbalch equation is defined as

$$pH = pKa + \log[\text{unionized/ionized}],$$

where Ka is the dissociation constant of the weak base in presence of water (i.e., higher the pKa, stronger is the base). Therefore for a weak base as the pH increases, the concentration of the unionized species also increases.

These equations facilitate the calculation of the relative concentrations of ionized (water-soluble) and unionized (lipid-soluble) species of a drug at any given pH, provided Ka of the drug is known.

Particle Size

Decreasing the particle size of drugs may enhance the rate of solubilization by increasing the effective surface area of the drug that is exposed to the dissolution medium. Therefore, for those drugs with limited aqueous solubility, particle size plays an important role in determining the rate and extent of drug absorption. Examples of some drugs that demonstrate varying drug absorption with particle size include griseofulvin, phenacetin, digoxin, and nitrofurantoin. The gastrointestinal absorption of ultramicrocrystalline griseofulvin, a fungistatic agent is reportedly 1.5-fold greater than that of the microsize griseofulvin. Increased absorption due to a reduction in the particle size allows for the administration of lower doses with the ultramicrocrystalline compound (500 mg vs. 375 mg, once daily for the microsize griseofulvin).

Physiological and Pathological Factors Affecting Drug Absorption

Physiological factors at the site of administration also influence the drug absorption from a given dosage form. Dissolution of drugs into the surrounding medium and permeation across biological membranes represent the primary factors guiding the rate and extent of drug absorption.

pH Profile of the Gastrointestinal Tract

The pH of the GI tract varies across its length, ranging from acidic (pH 1–3) in the stomach to alkaline (pH 7–8) in the large intestine. Drug absorption preferentially occurs in the slightly alkaline pH (6.5) of the small intestine. As explained earlier, most drugs are either weak acids or bases. The unionized or neutral form of the drug is preferentially absorbed across the lipophilic intestinal barriers. Depending on the pKa of the drug and the pH of the surrounding environment, the degree of unionized drug available for membrane transport varies. Therefore a small change in the pH of the environment is likely to cause a significant change in the percentage of unionized drug available for absorption. However, presence of even a small fraction of unionized drug is usually sufficient to allow absorption. This is because of the continual existence of equilibrium between the ionized and unionized forms that allows formation of more unionized drug whenever some of it is absorbed. In addition to its effect on ionization, the pH of the GI tract may also affect the absorption of a drug due to its potential influence on the stability and solubility of the drug.

Gastric Emptying Rate

Because most drugs are primarily absorbed from the small intestine, their absorption from the GI tract is influenced by the gastric emptying rate, which refers to the release of stomach contents into the first segment of the small intestine (i.e., duodenum). Increased or decreased gastric emptying may impact the stability, dissolution, as well as the rate and extent of absorption of a drug from the GI tract. For example, a direct correlation has been found between gastric emptying rate and systemic absorption of acetaminophen, a drug that has negligible absorption from the stomach. Prolonged residence of certain drugs like aspirin may irritate the stomach mucosa if the gastric emptying is delayed. For some drugs such as L-dopa, penicillin, the acidic pH of the stomach may result in loss of stability in the case where gastric emptying is prolonged. For some drugs with poor aqueous solubility, longer residence time in the stomach may aid in increased dissolution. Several factors have been shown to have varying influences on the gastric emptying rate: quality of ingested meals (volume, viscosity, temperature, pH, and composition), diurnal variations, age, presence of diseases or conditions (gastric ulcers, diabetes, peritoneal irritation, pregnancy, surgery, pain, and emotional stress), body posture, coadministration of alcohol, and drugs (anticholinergics, antihistamines, opioid analgesics, anesthetics, antacids, dopamine D2-receptor agonists, and antagonists).

Intestinal Transit Time

For maximal systemic absorption from the GI tract, a drug should reside at its intestinal site of absorption for an adequate amount of time before it undergoes elimination in the feces. Factors that influence the intestinal motility and thereby alter the transit time may therefore contribute to incomplete absorption of the drug from a dosage form. Presence of solid food in the small intestine delays the transit time, however, the net effect on absorption is variable due to the potential influences of the nutritional contents and viscosity of the food on drug dissolution at the site of absorption. Also, coadministered medications as well as presence of disease conditions such as dehydration or diarrhea may influence the net transit time and intestinal absorption of drugs from the GI tract.

Blood Flow to the Gastrointestinal Tract

Ample blood flow to the small intestine is one of the factors responsible for the faster absorption rates from this region. Factors that may influence the blood flow to the GI tract such as presence of food in the gut, cardiac disease, exercise, and so on may therefore have an impact on the net absorption of drugs.

Effect of Food on Drug Absorption

Presence of food in the GI tract may alter the rate and/or the extent of systemic absorption of a drug. The outcome of a food–drug interaction can be a delayed, reduced, or increased absorption and may depend on the physicochemical properties of the drug, the formulation excipients, and the degree to which these may influence the stability, disintegration, release, and dissolution of the drug in the GI tract. In addition, the volume and nutritional content of the coingested food as well as the time of food administration relative to drug intake also influence the food effect outcome due to their varying effect on the GI physiology. The largest food-effect is often observed with foods that are high in fat content and for this reason, the food-effect bioavailability and fed bioequivalence studies are often conducted in presence of high-calorie and high-fat meals.

Various intrinsic mechanisms may be involved in the observed effect of food on bioavailability of drugs, including postprandial changes in the gastro-intestinal motility, gastric pH, biliary secretions, and blood flow that alter the GI residence time, solubility, intestinal permeability, and, ultimately, the systemic bioavailability of drugs. On the other hand, certain foods may induce or inhibit certain drug metabolizing enzymes in the GI tract (e.g., grapefruit juice), thereby altering the extent of presystemic metabolism of coadministered drugs and their resultant systemic bioavailability. While, an understanding of the type and magnitude of food-effect on the systemic absorption of a drug is often adequate to enable optimal dosage recommendations, knowledge of the mechanisms leading to the observed effect may prove beneficial; for example, when attempting to address the food-effect of a novel or altered formulation.

Delayed drug absorption (i.e., a prolonged T_{max}) in the presence of food may not usually warrant dosage adjustments as long as the extent of absorption is unaffected. In the case of NSAIDs (ibuprofen, ketoprofen, flurbiprofen, naproxen, etc.) a delayed absorption can be useful to avoid GI side effects of these drugs. Reduced drug absorption (i.e., lower AUC_{inf}) in presence of food may require that the drug be taken on an empty stomach (e.g., trospium chloride, rifampin). However, in some cases when the extent of decrease is not significant in terms of clinical outcome, dosage recommendations may not be needed. Increased drug absorption in presence of food, may have varying implications depending on the therapeutic index of the drug and the clinical utility or potential risk of an increased systemic drug exposure. While in some drugs increased drug absorption with food may be utilized to achieve maximal systemic bioavailability (e.g., itraconazole, ritonavir, saquinavir), for drugs with narrow therapeutic indices concomitant food intake may not be recommended due to increased frequency of adverse events (e.g., efavirenz).

Presence of high-fat food in the GI tract may increase the bioavailability of certain poorly soluble drugs as a consequence of their enhanced solubility in bile (e.g., griseofulvin, troglitazone, halofrantine). In addition to meals, many beverages including water, milk, juices, carbonated drinks, alcohol as well as nutrients such as iron, magnesium, calcium, and certain vitamins have been shown to influence the GI absorption of drugs due to their varying effects on rates of drug dissolution, gastric emptying, gastric pH, stability of the drug, as well as presystemic metabolism.

Drug–Excipient Interactions

The choice of excipients for use with an active ingredient in a formulation may impact the stability, solubility as well as bioavailability of the drug. Most interactions result in decreased stability of the drug (e.g., ascorbic acid and penicillin G, bisulfite and epinephrine, edetate salts and insulin, thiomerosal, amphotericin, hydralazine HCl, meta-cresol and chlorpromazine, amlodipine, fluoxetine and lactose, etc.). The absolute bioavailability of ranitidine from an immediate release encapsuled pellet formulation containing the excipients PEG 400 was found to be significantly reduced (from 51% to 35%). PEG 400 also affected the absorption rate of ranitidine as demonstrated via major differences observed in the T_{max} and C_{max}, suggesting that it adversely influences the GI absorption of ranitidine.

Presystemic Metabolic Enzymes and Transporters

Peptide drugs, such as insulin and heparin, are usually indicated for chronic conditions, requiring daily injections for a long-term treatment. Gastric digestive enzymes, in addition to acidic environment and bacterial degradation present a major hurdle in oral delivery of peptide drug molecules. Contribution of drug metabolizing enzyme expression and activity in GI tract as a major component of the first pass metabolism of drugs has been a subject of interest over the years. Some enzymes, particularly Cytochrome P450 (CYP) enzymes, have been extensively studied more so than others. Sufficient evidence is available to indicate the major role of intestinal CYP3A4 and CYP1A2 in the metabolism of drugs even before reaching systemic circulation. Active transport mechanisms in GI tract, as discussed earlier, play an important role by themselves and also in conjunction with the drug metabolizing enzymes. For example, several researchers have noted the coexpression of drug metabolizing enzyme, CYP3A4, and drug efflux-transporter, P-glycoprotein, in the GI tract. Their co-expression is believed to limit the absorption of xenobiotics including a variety of drugs. Intestinal metabolism may be a major route of clearance for an orally administered CYP3A4 substrate (e.g., midazolam). Systemic absorption is limited due to P-gp mediated efflux for its substrates (e.g., digoxin). A combination of metabolism and efflux may limit the absorption of a CYP3A4 and P-gp substrate (e.g., cyclosporine).

Drug–Drug Interactions

Changes in drug absorption may occur because of physiological or physicochemical factors that are altered by precipitant drug. Some of the factors previously discussed, such as pH of gastric contents, gastric emptying time, presystemic metabolic enzymes, and transporters, might be altered due to the presence of another drug. Drugs that reduce gastric pH (antacids, H_2-antagonists, proton-pump inhibitors, etc.) might alter the solubility of a coadministered drug. Opioid drugs are known to cause delay in gastric emptying time. Depending on the physicochemical parameters, these physiological changes could result in changes in C_{max}, AUC or T_{max} of target drug. Oral clearance of drugs metabolized in intestine could be increased or decreased when coadministered with inhibitors or inducers of metabolic enzymes.

Drug–Disease Interaction

With regards to oral administration, GI diseases resulting in pH changes, gastric emptying time changes and, GI resection would potentially affect the absorption of drugs. However, the changes, to the distribution, and clearance of drugs due to celiac and inflammatory bowel diseases must be considered before attributing abnormal serum concentrations of drugs to malabsorption. GI disease may slow gastric emptying and delay the complete absorption of drugs when their rate of absorption depends on gastric emptying time. Other inflammatory GI diseases such as graft-versus-host disease (GVHD) of the gut, Behcet's syndrome, and scleroderma involving the GIT may directly reduce absorption of drugs such as acetaminophen, amitriptyline, certain anticonvulsants, benzodiazepines, and cyclosporine. Surgical resection drastically changes the anatomy of the GIT and alters important variables in the absorption process. Diminished small bowel surface area is shown to reduce the extent of absorption of phenytoin, digoxin, cyclosporin, aciclovir, hydrochlorothiazide, and certain oral contraceptives. The underlying cause of the reduction is unknown. When gastric emptying time or pH is altered by surgery, the rate of drug absorption appears to be reduced. However, it is not clear which variable is more important in determining therapeutic effects.

TOOLS FOR ASSESSING DRUG ABSORPTION

Use of drug absorption models early on in the drug development process increases the probability of success by identification of drug candidates with good absorption potential. Systemic bioavailability of drugs is a composite of processes that favor and inhibit absorption such as active/passive/facilitated uptake processes, removal via efflux transporters such as P-gp, degradation via drug metabolizing enzymes such as CYP450s. Therefore, early assessment of the relative contributions of each of these components to drug absorption is useful to eliminate potential "problem" drugs and identify those molecules that have good pharmacokinetic profile to go hand in hand with their biological effectiveness. Several in vitro and in vivo methods with a range of throughputs and predictabilities, as well as computational methods have been used with varying rates of success to predict drug absorption potential.

Cell Culture Models

The use of human colon adenocarcinoma-derived Caco-2 cell cultures as a model to mimic the morphological and functional characteristics of the mature enterocyte is well established and is routinely used for screening the intestinal permeability of discovery compounds. When grown on membrane filter supports, cultures differentiate into intestinal cell-like structures with morphological features such as microvilli, carrier-mediated transporters, tight junctions, efflux proteins, and enzymes. Drug transport from the donor compartment, across the monolayer-bearing membrane to the acceptor compartment is measured as the apparent permeability coefficient, P_{app}. While the physicochemical compatibility of the drugs for undergoing passive diffusion is understood by this method, due to the expression of efflux proteins such as P-gp and drug metabolizing enzymes such as CYP450s in these cells, the contribution of these processes to the overall drug delivery is also understood. Several improvements serve to enhance the in vitro prediction capabilities of Caco-2 cultures. These include novel high-throughput and automated techniques, isolation of more homogenous populations of clones to reduce heterogeneity, transfection of cells to improve the expression of specific transport proteins or drug metabolizing enzymes, and identification and standardization of culture-conditions to reduce variability in the morphology (growth media, type of membrane support, seeding density, etc.).

The Madin Darby canine kidney (MDCK) cell line is another model that mimics intestinal epithelium and is particularly suitable to evaluate the passive diffusion of drugs across membranes due to the low expression of transport proteins and drug metabolizing enzymes. The time for establishing a confluent culture and initiating the transport studies is also smaller (7–14 days) compared to Caco-2 cell line (21 days). In addition, MDR1-MDCK cells formed by the transfection of the MDCK cell line with the *mdr1*-gene are a useful model to screen potential P-gp substrates. Other in vitro cell cultures that mimic small intestine include TC7, LS180, 2/4/A1, HT29-18-C1, as well as cocultures of these cell lines with Caco-2.

Artificial Membranes

Parallel artificial membrane permeation assay or PAMPA is a high throughput tool with potential to screen compounds capable of undergoing transcellular passive diffusion. The methodology involves use of an artificial membrane that is formed by impregnating a mixture of lecithin and an inert organic solvent, into a hydrophobic filter material (2). The filter is placed in contact with a 96-well microtiter plate completely filled with aqueous buffer solutions and transport studies are initiated by the addition of drug solution on top of the filter plate, to determine its permeability across the artificial membrane. In addition, the effect of pH, phospholipid composition, and effect of surfactants, etc. on the membrane permeability of drugs can also be evaluated. Because this model is devoid of transporters, channels or efflux proteins, it is suitable for a high throughput screening of compounds with transcellular diffusion potential. Therefore, when a drug is classified to have "low" permeability by the PAMPA method, it still needs to be screened for paracellular or active transport, using other in vitro methodologies. Nevertheless, because of its potential to rapidly identify those molecules with "high" permeability under the conditions employed, it could be a very useful tool in early drug development.

Another high throughput technique for characterizing membrane partition coefficients of drugs is the use of IAM or immobilized artificial membrane chromatography. IAMs are chromatographic surfaces prepared by covalently immobilizing cell membrane phospholipids to solid surfaces at monolayer densities to mimic fluid cell membranes (3). The retention time of a solute on the IAM HPLC column is then measured to determine its membrane partition coefficient.

In Silico Methods

While cell culture-based evaluation of intestinal permeability requires actual synthesis of discovery drugs and is far too time-consuming to be adequately high throughput, in silico or virtual models can allow the absorption screening of thousands of computationally-designed drug-like molecules even before they are synthesized. Such models can predict the solubility and membrane permeability of lead molecules by estimating their physicochemical properties from available structural information alone. Molecules that have been prioritized by such methods can then be synthesized and carried further into drug development. Such computational models can be qualitative or quantitative in nature and are most suited for the screening of drugs absorbed by passive intestinal permeability and may not adequately predict the permeability of compounds that require active transport mechanisms.

Qualitative in silico models are developed by comparing the molecular descriptors of successful drug molecules to those of not-drug-like molecules and identify patterns in drug properties that may contribute to the observed differences. One such example is a model-generated "rule of five" by Lipinski et al. (4), that states that poor absorption or permeation of a drug is more likely when there are more than 5 H-bond donors, 10 H-bond acceptors, the molecular weight is greater than 500, and the calculated Log P is greater than 5 (4). Exceptions to this rule may occur when drugs are predominantly transported by active transport mechanisms.

Several quantitative models to predict intestinal drug absorption have also been developed (5–7). These models are derived from vast libraries of drugs and are based on the observed mathematical relationship between the structural features or structurally-derived descriptor parameters for these drugs and their percentage in vivo intestinal absorptions. An external data set containing drugs that are distinct from those used in the learning set is then used to validate these models to determine their reliability in estimating drug absorption potentials. Using such models, the intestinal absorption potential of new molecules and the effect of structural alterations on the intestinal permeability can be reliably estimated to aid in lead optimization.

While the earlier described models predict drug absorption based on the physicochemical properties of the drug and their interaction with the biological membranes, these do not consider the role of physiological factors of the GI tract such as the pH profile, gastric emptying time, intestinal transit time, and so on that could influence the net absorption of drugs. Physiological-based models of drug absorption have been developed that incorporate not only the physicochemical properties of the drug but also the known physiology of the

GI tract into the model, thereby improving the prediction of membrane permeability (8–10).

Other techniques available for studying drug absorption include the in vitro techniques such as the everted rat intestinal sac model, use of Ussing-chambers, as well as in situ and in vivo perfusion of rat intestines. Although these methods are useful for obtaining information on the behavior of a drug in the GI environment, each comes with its own limitations including the loss of viability of the various isolated intestinal preparations, as well as the difficulty of extrapolating results obtained in animal models to humans.

In vivo assessment of drug absorption in humans using GI perfusion techniques provides the most thorough information on the role of GI physiology including presence of transporters, efflux proteins, enzymes, as well as physicochemical properties of drugs on their net absorption from the small intestine, although this may not be amenable for a high throughput screening. Catheters can be inserted through oral or nasal cavities of subjects under local anesthesia, for the perfusion of an open or isolated intestinal segment such as jejunum to study drug absorption, permeability and excretion of parent drug, and metabolites into the intestinal lumen. The regional intestinal perfusion is a closed segment approach that involves insertion into the desired region of the intestine, of a perfusion tube equipped with two occluding balloons, which when inflated generate an independent, perfusable intestinal segment. The intestinal segment is then perfused with the drug solution and an intravenous dose of the same drug is administered either simultaneously (using radio-labeled drug) or sequentially, following the perfusion experiment. By comparing the area under the plasma concentration-time curves of parent drug and metabolites for the intraluminal and intravenously administered drugs, it is possible to calculate the GI extraction of the drug (11–14).

Knowledge of specific factors affecting drug absorption is pivotal during drug development for making an appropriate choice of route of administration and optimal dosage form. The factors affecting the physiological functions and pathological conditions in humans and physicochemical properties of drugs together determine the fate of the drug. While it is difficult to prescribe the use of a single in vitro technique to predict in vivo drug absorption, use of several models is sure to improve our understanding of the absorption characteristics of a drug, early in development.

REFERENCES

1. Kim RB. Organic anion-transporting polypeptide (OATP) transporter family and drug disposition. Eur J Clin Invest 2003; 33(suppl)2:1–5.
2. Kansy M, Senner F, Gubernator K. Physicochemical high throughput screening: parallel artificial membrane permeation assay in the description of passive absorption processes. J Med Chem 1998; 41:1007–1010.
3. Pidgeon C, Cai SJ, Bernal C. Mobile phase effects on membrane protein elution during immobilized artificial membrane chromatography. J Chromatogr A 1996; 721:213–230.

4. Lipinski CA, Lombardo F, Dominy BW, Feeney PJ. Experimental and computational approaches to estimate solubility and permeability in drug discovery and development settings. Adv Drug Deliv Rev 2001; 46:3–26.
5. Klopman G, Stefan LR, Saiakhov RD. ADME evaluation. 2. A computer model for the prediction of intestinal absorption in humans. Eur J Pharm Sci 2002; 17:253–263.
6. Wessel MD, Jurs PC, Tolan JW, Muskal SM. Prediction of human intestinal absorption of drug compounds from molecular structure. J Chem Inf Comput Sci 1998; 38:726–735.
7. Zhao YH, Le J, Abraham MH, et al. Evaluation of human intestinal absorption data and subsequent derivation of a quantitative structure-activity relationship (QSAR) with the Abraham descriptors. J Pharm Sci 2001; 90:749–784.
8. Dressman JB, Fleisher D, Amidon GL. Physicochemical model for dose-dependent drug absorption. J Pharm Sci 1984; 73:1274–1279.
9. Willmann S, Schmitt W, Keldenich J, Dressman JB. A physiologic model for simulating gastrointestinal flow and drug absorption in rats. Pharm Res 2003; 20:1766–1771.
10. Yu LX, Amidon GL. A compartmental absorption and transit model for estimating oral drug absorption. Int J Pharm 1999; 186:119–125.
11. Knutson L, Odlind B, Hallgren R. A new technique for segmental jejunal perfusion in man. Am J Gastroenterol 1989; 84:1278–1284.
12. Lennernas H, Ahrenstedt O, Hallgren R, Knutson L, Ryde M, Paalzow LK. Regional jejunal perfusion, a new in vivo approach to study oral drug absorption in man. Pharm Res 1992; 9:1243–1251.
13. Palm K, Luthman K, Ungell AL, Strandlund G, Artursson P. Correlation of drug absorption with molecular surface properties. J Pharm Sci 1996; 85:32–39.
14. von RO, Greiner B, Fromm MF, et al. Determination of in vivo absorption, metabolism, and transport of drugs by the human intestinal wall and liver with a novel perfusion technique. Clin Pharmacol Ther 2001; 70:217–227.

3

Approaches to Developing
In Vitro–In Vivo Correlation Models

Adrian Dunne

School of Mathematical Sciences,
University College Dublin, Dublin, Ireland

INTRODUCTION

The process of establishing a Level A in vitro–in vivo correlation (IVIVC) consists of a number of steps. Those steps associated with the modeling and data analysis can be described as follows.

1. Construct a model describing the functional relationships (1) between the quantities of interest.
2. Connect this model with the data by specifying the statistical aspects of the data, and thereby developing a structural model (1).
3. Estimate the model parameters from the data (model fitting).

The literature on Level A IVIVC modeling and data analysis is quite extensive, and a wide variety of models and data analytic approaches have been reported. It can be difficult to compare many of these models and methods to see what they have in common and how they differ. This difficulty arises because the aforementioned steps are not always clearly delineated or described. The fact that these models look quite different also makes comparisons difficult. The objective of this chapter is to separate these steps and consider the first two of them in relation to some of the methods and models that have been described in the literature. Some models which appear quite different will be shown to be equivalent in certain respects. The importance of the second step in describing the statistical properties of the data will be emphasized. The third of the steps mentioned above will be commented upon briefly.

THE FUNCTIONAL RELATIONSHIP

Introduction

The first step in the process of developing an IVIVC model involves the construction of a model describing the functional relationships of interest. This model is deterministic in that it does not take into account the random element of the data used to establish the IVIVC. In fact, it does not deal with the data at all. The objective is simply to describe the relationships between the quantities of interest without considering the processes of collecting samples and conducting assays, and so on, which give rise to the data. The connection between the quantities involved in the functional model and the data collected will be considered in the section "The Structural Model." The aim of this section is to describe a modeling framework to which many of the models reported in the literature belong, and thus facilitate their comparison and, perhaps, the development of future models.

The framework that will be employed is based on dynamic systems analysis (2). A linear time invariant dynamic system with a single input and single output is shown schematically in Figure 1. Such a system has an input $i(t)$, which may be changing with time, and an output $o(t)$, which is also a function of time. It is important to define the system being considered by specifying its boundary. Everything inside the boundary is considered to be

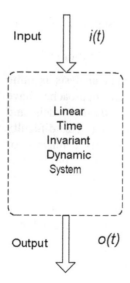

Figure 1 A schematic diagram showing a linear time invariant dynamic system with input $i(t)$ and output $o(t)$.

part of the system and that which is outside the boundary is the environment. Describing the system as linear means that the superposition principle applies, that is, the output corresponding to an input $i_1(t) + i_2(t)$ is the sum of the outputs corresponding to $i_1(t)$ and $i_2(t)$ applied separately. Time invariance implies that shifting the input in time produces a similar shift in the output, that is, the input $i(t + \Delta)$ has an output $o(t + \Delta)$. Note that there is no assumption that the processes giving rise to the input constitute a linear time invariant system. From here on, unless otherwise noted, all systems will be linear time invariant dynamic (LTID) systems and will be referred to simply as LTID systems for ease of description. An LTID system can be characterized by its response to any standard input, because knowing the output corresponding to the standard input facilitates prediction of the output corresponding to any other input. The standard input most commonly used is the unit impulse $\delta(t)$, that is, a pulse of unit area and zero duration. The corresponding output $r(t)$ is characteristic of the system and is known as the unit impulse response. Knowing the unit impulse response allows us to predict the response to any other input using the following argument: Any input to an LTID system can be viewed as the sum of a series of impulses of different magnitudes each shifted slightly in time from the previous one. Consequently, the output would be the sum of unit impulse responses adjusted for the magnitude of the input and shifted in time by an amount corresponding with the input. In the limit, this sum becomes an integral allowing us to write

$$o(t) = \int_0^t i(\tau)r(t - \tau)\,d\tau \qquad (1)$$

This integral, which is known as the convolution integral (2), expresses the relationship between the system characteristics ($r(t)$), the input ($i(t)$), and the corresponding output ($o(t)$).

Framework for In Vitro–In Vivo Correlation Model

IVIVC modeling aims to use information about in vitro dissolution of a solid dosage form to predict its in vivo performance. In order to achieve this, a relationship is constructed between in vitro and in vivo dissolution of drug from the dosage form. This relationship is then extended such that it can be used to predict the in vivo performance of the dosage form. The general framework for the model is outlined in this section.

The fraction (or percentage) of the dose dissolved in vitro under specific controlled conditions, as a function of time will be represented by $F_1(\theta_1,t)$ where the vector θ_1 consists of the parameters of this function and t is time elapsed since the dosage unit was introduced into the dissolution medium. The rate of in vitro drug dissolution at any time is the derivative of this function with respect to time and will be denoted by $F_1'(\theta_1,t)$ where the prime ($'$) indicates differentiation with respect to time, that is,

$$F_1'(\theta_1,t) = \frac{dF_1(\theta_1,t)}{dt} \tag{2}$$

Similarly, $F_2(\theta_2,t)$ will be used to represent the fraction of drug dissolved in vivo as a function of time. The vector θ_2 represents the parameters of this function, and t is time elapsed since the dosage unit was administered. Using the same convention as that adopted earlier, the rate of drug dissolution in vivo will be written as $F_2'(\theta_2,t)$.

Since it is desired to construct a relationship between in vitro and in vivo dissolution of drug from the dosage form, the fraction (or percentage) of the dose dissolved in vivo is the quantity of interest. However, in practice, it is difficult to observe this directly and instead the plasma drug concentration is the quantity observed. This may in fact be advantageous because the purpose of the IVIVC model is to predict the in vivo performance of the dosage form and this is often related to plasma drug concentration, for example, area under the curve (AUC) or C_{max} (3,4). Plasma drug concentration is linked to in vivo drug dissolution via drug absorption and is also affected by drug distribution, metabolism, and elimination. Putting this into a dynamic systems framework, some or all of these processes will be part of the LTID system and the plasma drug concentration will be considered as the output. The input depends on how we define the boundary of the system, but will depend in some way on the in vivo drug dissolution. Some alternative definitions of the system boundary are discussed subsequently. The LTID system provides the connection between plasma drug concentration and in vivo drug dissolution, which is linked to in vitro drug dissolution via the in vitro–in vivo model. Consequently, using the convolution integral we can write

$$C(t) = \int_0^t i(\tau)r(t-\tau)d\tau \tag{3}$$

where $C(t)$ denotes the plasma drug concentration as a function of time, $i(t)$ is the input into the system, and $r(t)$ is the unit impulse response. The input function $i(t)$ describes how the rate at which drug enters the system changes with time.

The unit impulse response characterizes the LTID system and is the link between the input and the output, as is clear from Equation 3. It may be written as

$$r(t) = f_3(\theta_3,t) \tag{4}$$

where $f_3(\cdot)$ is a function with parameter vector θ_3. Information on this function is frequently gathered by inputting an impulse into the system and studying the corresponding output. This impulse, which is known as the "reference," may be an intravenous bolus (5), an oral solution (6), or an immediate release (IR) dosage form (7). When the reference is an intravenous bolus, the LTID system does not include drug absorption as part of the system because the reference bypasses the absorption mechanisms by being delivered directly into the blood. As a result, drug absorption becomes part of the input into the system. On the other hand, a reference consisting of an oral solution or an IR dosage form does undergo absorption into the blood and, consequently, these absorption processes are part of the LTID system being characterized by the reference. In addition, an IR dosage form does have to dissolve and although the label "IR" implies that this is instantaneous, in practice it does take a finite amount of time and this also is part of the LTID system. Instead of administering a reference dose to characterize the system, it is possible to assume a particular form for $r(t)$, such as a compartmental model (8,9).

The input into the system depends in some way on the in vivo drug dissolution and is therefore the link between drug dissolution and the plasma drug concentration. This dependence may be expressed as

$$i(t) = f_4(\theta_4, t, F_2(\theta_2, t)) \tag{5}$$

where $f_4(\cdot)$ is a function with parameter vector θ_4.

Finally, the relationship between in vitro and in vivo dissolution must be modeled

$$F_2(\theta_2, t) = f_5(\theta_5, t, F_1(\theta_1, t)) \tag{6}$$

where $f_5(\cdot)$ is also a function with parameter vector θ_5. Equation 6 is at the heart of IVIVC modeling because it expresses the relationship between in vitro and in vivo drug dissolution. Of course, Equations 5 and 6 could be combined to express the system input in terms of the fraction of the dose dissolved in vitro.

This so-called convolution model or convolution approach to in vivo–in vitro modeling has been used by many researchers in this area (7,8,10,11).

Examples

The Unit Impulse Response

Consider the case where the unit impulse response function is given by

$$r(t) = \frac{1}{V} \exp(-k_e t) \tag{7}$$

This is the appropriate form for the function $r(t)$ when the LTID system consists of a one-compartment model with first-order elimination. The apparent volume of distribution is denoted by V, and the first-order elimination rate constant by k_e. Such a model is depicted in Figure 2. Note that drug absorption is not part of the LTID system in this case and, as a consequence, is not assumed to be linear. Using this unit impulse response in the convolution integral in Equation 1 gives

$$C(t) = \frac{1}{V} \int_0^t i(\tau) \exp(-k_e(t - \tau)) d\tau \tag{8}$$

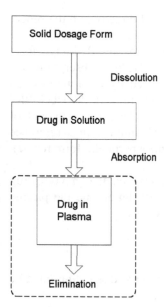

Figure 2   A one-compartment pharmacokinetic model. The *dashed box* outlines the linear time invariant dynamic (LTID) system, which does not include drug absorption.

Differentiating both sides of Equation 8 with respect to t gives

$$V\frac{dC(t)}{dt} = i(t) - k_e VC(t), \tag{9}$$

which demonstrates that the model based on the convolution integral in Equation 8 is identical to that based on the differential Equation 9 (12,13). Rearranging Equation 9 and integrating both sides from 0 to t gives

$$\int_0^t i(\tau)d\tau = VC(t) + k_e V\int_0^t C(\tau)d\tau, \tag{10}$$

which is the Wagner–Nelson equation (14). This demonstrates that the model based on the convolution integral in Equation 8 is also identical to that based on a Wagner–Nelson deconvolution. Hence, any model based on a Wagner–Nelson deconvolution is effectively using Equation 8 or its equivalent, Equation 9.

Now, consider the case where the unit impulse response is given by

$$r(t) = \frac{A}{V}\exp(-\lambda_1 t) + \frac{1-A}{V}\exp(-\lambda_2 t) \tag{11}$$

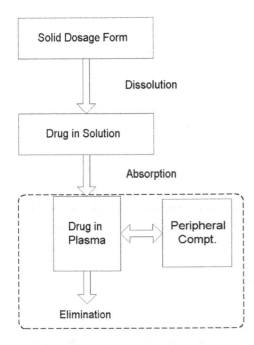

Figure 3 A two-compartment pharmacokinetic model. The *dashed box* outlines the linear time invariant dynamic (LTID) system, which does not include drug absorption.

The two-compartment model with elimination from the central compartment illustrated schematically in Figure 3 has such a unit impulse response function with

$$\lambda_1 + \lambda_2 = k_e + k_{12} + k_{21} \quad \lambda_1\lambda_2 = k_e k_{21}$$

$$A = \frac{k_{21} - \lambda_1}{\lambda_2 - \lambda_1} \tag{12}$$

where k_{12} and k_{21} are the first-order rate constants for transfer from the central to the peripheral compartment and vice versa, respectively. The first-order elimination rate constant and the apparent volume of distribution of the central compartment are denoted by k_e and V, respectively. As in the previous example, this LTID system does not include drug absorption. The corresponding convolution integral is

$$C(t) = \int_0^t i(\tau)\left[\frac{A}{V}\exp(-\lambda_1(t-\tau)) + \frac{1-A}{V}\exp(-\lambda_2(t-\tau))\right]d\tau \tag{13}$$

If the amount of drug in the peripheral compartment at time t (denoted by $Q_p(t)$) is considered as the output of the system instead of $C(t)$, then

$$Q_p(t) = \int_0^t i(\tau)[B\exp(-\lambda_1(t-\tau)) - B\exp(-\lambda_2(t-\tau))]d\tau \tag{14}$$

where

$$B = \frac{k_{12}}{\lambda_2 - \lambda_1} \tag{15}$$

and λ_1 and λ_2 are as defined in Equation 12. Differentiation of Equation 13 with respect to t and using Equation 14 in simplifying gives

$$V\frac{dC(t)}{dt} = i(t) + k_{21}Q_p(t) - (k_e + k_{12})VC(t) \tag{16}$$

This again shows that the model based on the convolution integral is identical to that based on the differential equation. Equation 16 can be rearranged and integrated from zero to t to give

$$\int_0^t i(\tau)d\tau = VC(t) + k_eV\int_0^t C(\tau)d\tau + Q_p(t), \tag{17}$$

which is the Loo–Riegelman equation (15). This shows that the model based on the convolution integral is identical to that based on a Loo–Riegelman deconvolution. Hence, any model based on a Loo–Riegelman deconvolution is effectively using Equation 13 or its equivalent, the differential Equation 16.

If drug absorption is included as part of the LTID system as depicted in Figure 4 and is first order, then the one-compartment model has a unit

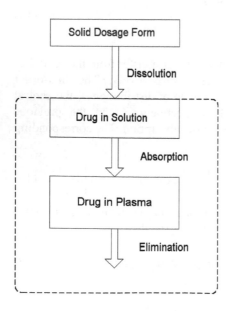

Figure 4 A one-compartment pharmacokinetic model. The *dashed box* outlines the linear time invariant dynamic (LTID) system, which includes drug absorption.

impulse response function given by

$$r(t) = \frac{k_a}{V(k_a - k_e)}[\exp(-k_e t) - \exp(-k_a t)] \tag{18}$$

where V and k_e are as described earlier and k_a is the first-order absorption rate constant. The corresponding convolution integral is

$$C(t) = \frac{k_a}{V(k_a - k_e)} \int_0^t i(\tau)[\exp(-k_e(t - \tau)) - \exp(-k_a(t - \tau))] d\tau \tag{19}$$

Differentiating both sides of Equation 19 with respect to time yields the differential equation

$$V\frac{dC(t)}{dt} = k_a Q_a(t) - k_e V C(t) \tag{20}$$

where $Q_a(t)$ is the quantity of drug available for absorption at time t. It can also be shown that

$$\frac{dQ_a(t)}{dt} = i(t) - k_a Q_a(t) \tag{21}$$

Hence, the model based on the convolution integral is equivalent to that based on the differential Equations 20 and 21.

When drug absorption is included as part of the LTID system and is first order, then the two-compartment model has a unit impulse response function given by

$$r(t) = \frac{k_a}{V}\left[\frac{k_{21} - k_a}{(\lambda_1 - k_a)(\lambda_2 - k_a)}\exp(-k_a t) + \frac{k_{21} - \lambda_1}{(k_a - \lambda_1)(\lambda_2 - \lambda_1)}\exp(-\lambda_1 t)\right.$$

$$\left. + \frac{k_{21} - \lambda_2}{(k_a - \lambda_2)(\lambda_1 - \lambda_2)}\exp(-\lambda_2 t)\right] \tag{22}$$

with all parameters as previously defined. This LTID system is shown in Figure 5. The corresponding convolution integral is

$$C(t) = \int_0^t i(\tau)\frac{k_a}{V}\left[\frac{k_{21} - k_a}{(\lambda_1 - k_a)(\lambda_2 - k_a)}\exp(-k_a(t - \tau))\right.$$

$$+ \frac{k_{21} - \lambda_1}{(k_a - \lambda_1)(\lambda_2 - \lambda_1)}\exp(-\lambda_1(t - \tau))$$

$$\left. + \frac{k_{21} - \lambda_2}{(k_a - \lambda_2)(\lambda_1 - \lambda_2)}\exp(-\lambda_2(t - \tau))\right] d\tau \tag{23}$$

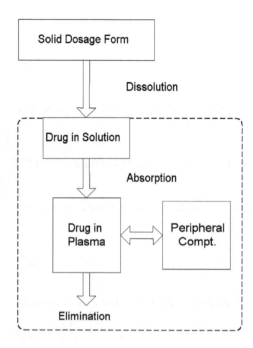

Figure 5 A two-compartment phar-macokinetic model. The *dashed box* outlines the linear time invariant dynamic (LTID) system, which includes drug absorption.

Following differentiation with respect to time, this gives

$$V\frac{dC(t)}{dt} = k_a Q_a(t) + k_{21} Q_p(t) - (k_e + k_{12})VC(t) \tag{24}$$

It can also be shown that Equation 21 holds in this case too and, consequently, the model based on the convolution integral in Equation 23 is equivalent to that described by the differential Equations 24 and 21.

All of the cases considered above are special cases of the poly-exponential model for the unit impulse response (8,11).

The next case to be considered is where the model for the unit impulse response $r(t)$ is saturated, that is, it contains a parameter for every observation time. Note that the word "saturated" is used here in the statistical sense and does not relate to saturable pharmacokinetic processes. In this case

$$r(t_i) = \theta_{3i} \quad i = 1, 2, \ldots \tag{25}$$

where t_i represents the ith observation time. Note that this model does not define the unit impulse response function at any time other than the observation times and consequently if the function is required to be evaluated at other times (as would be necessary to evaluate the convolution integral) then interpolation and possibly extrapolation will have to be employed. Any method based on numerical deconvolution (5,6,16) effectively uses this model. It has the advantage that no assumption is required about the structure of the LTID system and as a result the only way in which this part of the model can be mis-specified is through an inappropriate choice of the interpolation and/or extrapolation function. The disadvantages are that the model has more parameters than other models and that a reference dose is essential with this model because no assumption is being made about the structure of the LTID system and therefore the unit impulse response must be studied directly. Drug absorption may or may not be included as part of the LTID system, depending on the type of reference used as discussed earlier.

The Input Function

Any model that does not include drug absorption as part of the LTID system has the combined processes of in vivo dissolution and absorption as its input function. Such a model does not necessarily assume that drug absorption is linear or time invariant. Consider, for example, the model used by Polli et al. (17). This model included a Wagner–Nelson deconvolution, hence it was based on a one-compartment model with first-order elimination as described in Figure 2 and Equations 8, 9, and 10. The input into the system is absorption, which is preceded by dissolution. Although it was not necessary to do so, Polli et al. (17) assumed that both in vivo dissolution and absorption were first order, which gives the following input function

$$i(t) = \frac{k_a k_d f_a D}{k_a - k_d} [\exp(-k_d t) - \exp(-k_a t)] \tag{26}$$

where k_d and k_a are the first-order rate constants for dissolution and absorption, respectively, f_a is the bioavailability, and D is the dose administered. This input function is obviously dependent on both dissolution and absorption. If one assumes that absorption is so much faster than dissolution that absorption is effectively instantaneous (18), then $k_a \gg k_d$ and Equation 26 reduces to

$$i(t) \approx k_d f_a D \exp(-k_d t) = F_2'(\theta_2, t), \tag{27}$$

which reflects drug dissolution only.

Models in which absorption is part of another LTID system have been proposed (18) and in this case the output from that system becomes the input into the LTID system, which has the plasma drug concentration as its output. Assuming first-order absorption

$$i(t) = k_a \int_0^t F_2'(\theta_2, \tau) \exp(-k_a(t - \tau)) d\tau \tag{28}$$

If one was to add the assumption that dissolution was itself a first-order process, this equation would be identical to Equation 26.

Some models do include the absorption as part of the LTID system, as illustrated for the one-compartment and two-compartment models in Figures 4 and 5, respectively. In this case, the absorption process is assumed to be linear and time invariant. Models which use an oral drug solution or IR dosage form as a reference belong to this class (7). In these models, the input function is the rate of drug dissolution in vivo since absorption is part of the LTID system, in which case

$$i(t) = F_2'(\theta_2, t) \tag{29}$$

In Vitro Dissolution

Many different models have been used to describe the process of in vitro dissolution (6,13,16,19–23), a selection of which is listed in Table 1. Some of these models have alternative versions and, consequently, the version given in Table 1 may differ somewhat from that appearing in some literature reports. It should be noted that a first-order model for in vitro dissolution is equivalent to the exponential model in Table 1. With the exception of the quadratic model, all of the models in Table 1 have the property that $F_1(\theta_1, 0) = 0$ and $F_1(\theta_1, \infty) = 1$, that is, they predict no dissolution at time zero and complete dissolution after a sufficiently long time has passed. These models can easily be modified to account for incomplete dissolution by scaling the model by another parameter representing the degree of dissolution. For example, the Weibull model would become

$$F_1(\theta_1, t) = \theta_{13}\left(1 - \exp\left(-\theta_{11}t^{\theta_{12}}\right)\right) \tag{30}$$

where $\theta_{13} \leq 1$ is the extent of dissolution of drug in the dosage unit.

One of the most commonly used approaches (5,9,16,24) does not use any explicitly stated model, which is equivalent to using a saturated model defining the function at the sampling times using a separate parameter for each time that is,

$$F_1(\theta_1, t_i) = \theta_{1i} \quad i = 1, 2, 3 \ldots \tag{31}$$

where t_i represents one of the observation times (7). Note that this model does not describe the fraction dissolved in vitro at any time other than the observation times. If the fractions dissolved in vitro at times other than the observation times are

Table 1 Some of the Functions that Have Been Used to Describe Drug Dissolution In Vitro

Description	Function
Exponential	$F_1(\theta_1, t) = 1 - \exp(-\theta_{11}t)$
Weibull	$F_1(\theta_1, t) = 1 - \exp(-\theta_{11}t^{\theta_{12}})$
Logistic	$F_1(\theta_1, t) = \dfrac{\exp(\theta_{11} + \theta_{12}\log(t))}{1 + \exp(\theta_{11} + \theta_{12}\log(t))}$
Quadratic	$F_1(\theta_1, t) = \theta_{11} + \theta_{12}t + \theta_{13}t^2$
Hill	$F_1(\theta_1, t) = \dfrac{t^{\theta_{11}}}{\theta_{12} + t^{\theta_{11}}}$

required, then interpolation and extrapolation must be used. Like all saturated models, this model has the advantage that no assumption is required about the dissolution process and apart from the interpolation and/or extrapolation the model cannot be mis-specified. However, model mis-specification is rarely a problem with in vitro dissolution data and this advantage is probably more of theoretical than practical interest. The disadvantage is that of having more parameters than other models.

In Vivo–In Vitro Model

This part of the model relates the in vivo drug dissolution to the in vitro dissolution. The plasma concentration depends on the input function as is clear from the convolution integral and the input function depends on the in vivo dissolution, hence the in vivo–in vitro model completes the chain that links in vitro dissolution to plasma drug concentration. Many different models have been proposed, the simplest of which is the identity model

$$F_2(\theta_2, t) = F_1(\theta_1, t), \tag{32}$$

which has been widely used (5,12,17–19,25,26). Probably the most commonly used (6,8,16,27) is the "linear" model

$$F_2(\theta_2, t) = \theta_{51} + \theta_{52} F_1(\theta_1, t) \tag{33}$$

Nonlinear models have been used by various authors (6,28,29) and these have included Sigmoid, Hixon–Crowell, Gompertz, Weibull, Higuchi, Mitcherlich, and Logistic. These models are detailed in Table 2. Some of these models have several different versions and, consequently, the version given in Table 2 may not be exactly the same as some of those reported in the literature. One characteristic shared by all of these models is that the relationship between in vitro and in vivo drug dissolution is time invariant. This implies that this relationship is the same for slowly dissolving dosage forms as it is for rapidly dissolving

Table 2 Some of the Nonlinear Functions that Have Been Used to Describe the Relationship Between In Vivo and In Vitro Drug Dissolution

Description	Function
Sigmoid	$F_2(\theta_2, t) = \theta_{51} + \dfrac{\theta_{52} F_1(\theta_1, t)^{\theta_{53}}}{\theta_{54} + F_1(\theta_1, t)^{\theta_{53}}}$
Hixon–Crowell	$F_2(\theta_2, t) = \theta_{51} - (\theta_{51}^{0.33} - \theta_{52} F_1(\theta_1, t))^3$
Gompertz	$F_2(\theta_2, t) = \theta_{51} \exp(-\theta_{52} \exp(-\theta_{53} F_1(\theta_1, t)))$
Weibull	$F_2(\theta_2, t) = \theta_{51} - \theta_{52} \exp(-\theta_{53}(F_1(\theta_1, t))^{\theta_{54}})$
Higuchi	$F_2(\theta_2, t) = (\theta_{51} F_1(\theta_1, t))^{0.5}$
Mitcherlish	$F_2(\theta_2, t) = \theta_{51} - \theta_{52} \exp(-\theta_{53} F_1(\theta_1, t))$
Logistic	$F_2(\theta_2, t) = \dfrac{\theta_{51}}{1 + \theta_{52} \exp(-\theta_{53} F_1(\theta_1, t))}$

dosage forms. This is a potentially severe limitation of these models because it would not be surprising to find that the in vivo–in vitro relationship varies as the dosage form passes through the changing environment in the gastrointestinal tract. In other words, that the in vivo–in vitro relationship changes with time.

Models describing the in vivo drug dissolution rate as a time-dependent attenuation of the in vitro drug dissolution rate have been suggested. An exponential attenuation model (18) gives

$$F_2'(\theta_2, t) = \exp(-\theta_{51}t)F_1'(\theta_1, t) \tag{34}$$

A step function attenuation (frequently referred to as an "absorption window") was considered (8,13) as follows:

$$\begin{aligned} F_2'(\theta_2, t) &= F_1'(\theta_1, t) & 0 \le t \le T \\ &= 0 & \text{otherwise} \end{aligned} \tag{35}$$

and a sigmoid attenuation with

$$F_2'(\theta_2, t) = \frac{\exp(-\theta_{51}(t - T))}{1 + \exp(-\theta_{51}(t - T))}F_1'(\theta_1, t) \tag{36}$$

has been reported (13). A "Michaelis-Menten" type attenuation given by

$$F_2'(\theta_2, t) = \frac{\theta_{51} + F_1'(\theta_1, t)}{\theta_{52} + F_1'(\theta_1, t)}F_1'(\theta_1, t) \tag{37}$$

has also been proposed (8).

Time shifting and scaling is another common approach that can be used to describe the in vivo–in vitro relationship (8,9,11,13,22). Such models relate $F_2(\theta_2, t)$ to $F_1(\theta_1, \theta_{51} + \theta_{52}t)$ using linear or nonlinear models, alternatively they relate the derivatives of these two functions with respect to time. A nonlinear time scaling has also been reported (24).

A convolution model relating in vivo and in vitro drug dissolution was used by Veng-Pedersen et al. (11). It is not clear whether this approach is based on a purely empirical model or whether there is some other justification for it.

One of the problems with many of these models is that they could predict fractions of drug dose dissolved in vivo outside the range from 0 to 1, which, of course, is not possible. In order to overcome this, models based on transformations of $F_1(\theta_1, t)$ and $F_2(\theta_2, t)$, which are not constrained to the [0, 1] interval, have been proposed (7,30,31). These use the so-called link functions, such as the logit, log–log and complementary log–log, to transform fractions from the [0, 1] interval onto the interval $[-\infty, +\infty]$. This ensures that the back-transformed model predictions always lie in the [0, 1] interval.

THE STRUCTURAL MODEL

This step links the data collected to the functional model described earlier and in doing so it defines the statistical model for the data. It is an essential step in the whole process of establishing an IVIVC model because it forms the basis for the parameter estimation or model fitting. This is also the step which is most commonly ignored or at best, scant attention is paid to it.

Consider, for example, the plasma drug concentration at a particular time previously denoted by $C(t)$. The data set contains a value $Y(t)$, which is not equal to $C(t)$ but is an estimate of it. There are many reasons why $Y(t)$ and $C(t)$ are not equal, one of which is measurement error in the assay used to produce $Y(t)$. The difference between $Y(t)$ and $C(t)$ is not entirely predictable and is considered to be random. As a result, $Y(t)$ is a random variable whose statistical properties must be defined and it is this process that turns the functional model for the quantities of interest into a structural model for the data collected. The same is true for the in vitro dissolution data. The first of the statistical properties in question is the shape of the probability distribution. Because of its mathematical properties, the normal distribution is most commonly used for the probability distribution associated with the data. The other properties are the mean, variance, and correlation structure of the data.

The first choice that has to be made is whether or not the data will initially be transformed in some way and modeled as such and if so what are the appropriate transformations. Consider the plasma drug concentration as an example again. A simple model says that

$$Y(t) = C(t) + \varepsilon_1 \quad \varepsilon_1 \sim N(0, \sigma_1^2) \tag{38}$$

The ε_1 is known as the "error" term and represents the (random) difference between $Y(t)$ and $C(t)$. Equation 38 means that $Y(t)$ is normally distributed with mean $C(t)$ and variance σ_1^2, which is constant (does not change as t changes). Constant variance is described as homoscedasticity and non-constant variance as heteroscedasticity. Plasma drug concentration data must be positive and cannot therefore be normally distributed as in Equation 38, because the normal distribution includes negative values. One way around this is to assume that the (natural) logarithm of the measured plasma concentration is normally distributed as follows:

$$\ln(Y(t)) = \ln(C(t)) + \varepsilon_2 \quad \varepsilon_2 \sim N(0, \sigma_2^2) \tag{39}$$

This equation also states that the mean of the random variable $\ln(Y(t))$ is $\ln(C(t))$ and that it is homoscedastic, with variance denoted by σ_2^2. If one assumes that the error terms were zero, then Equations 38 and 39 would be equivalent. Zero error terms would mean that the measurements of plasma drug concentration were perfect, which is not a realistic assumption. With nonzero error terms, Equations 38 and 39 are not equivalent. This is because the logarithm of the mean value of a random variable is not equal to the mean value of the logarithm of the random variable and if the variance of a random variable is constant, then the variance of its logarithm is not constant and vice versa. These properties can be illustrated with a simple example where

$$C(t) = A \exp(-\alpha t) \tag{40}$$

with A and α taking values of 100 and 0.07, respectively. Let Equation 39 be true with σ_2^2 having a value of unity and the observations being uncorrelated with each other. Using the above model and parameter values, one can show, using simulation studies, that the mean value of $Y(t)$ is greater than $C(t)$, as illustrated in Figure 6. Furthermore, the same simulation studies demonstrate that the variance of $Y(t)$ is not constant, as illustrated in Figure 7. These results hold, in general, not just for this simple example and arise from the nonlinearity of the log transformation, and similar results would be expected with any nonlinear transformation. This example clearly demonstrates that the same functional

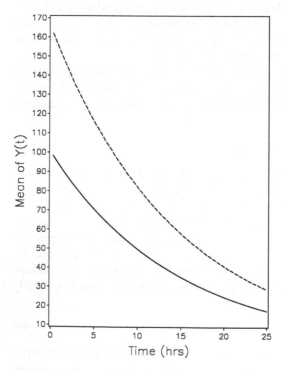

Figure 6   Results of the simulation study based on Equation (39). The *solid line* shows how the actual plasma drug concentration ($C(t)$) changes with time, and the *dashed line* shows how the mean value of the measurements of plasma drug concentration ($Y(t)$) changes with time when $\ln(Y(t))$ is normally distributed with mean $\ln(C(t))$. This figure demonstrates the difference between the values of $C(t)$ and the mean of $Y(t)$ based on the model in Equation 39.

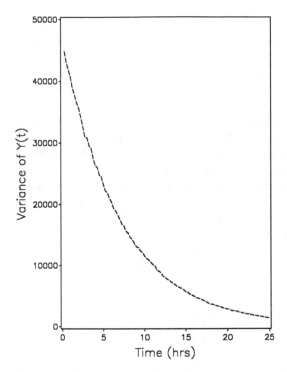

Figure 7 The *dashed line* shows how the variance of $Y(t)$ changes with time when the variance of $\ln(Y(t))$ is constant. The coefficient of variation of $Y(t)$ is constant as shown in Equation 41.

model can lead to different structural models, depending on the assumptions made in linking the data to the functional model. Hence, the process of constructing the structural model from the functional model is by no means trivial. Using different structural models in analyzing the data would, in general, give different results, that is, different estimates of the parameters of interest and different predictions. It is of course possible (in theory at least) to construct models for transformed and untransformed data, which are equivalent. For instance, in the case of the logarithmic transformation given in Equation 39, the equivalent model for $Y(t)$ is that it has a lognormal distribution, with mean and variance given by (32)

$$\text{Mean} = C(t)\exp{(0.5\sigma_2^2)} \quad \text{Variance} = C(t)^2 \exp{(\sigma_2^2)}(\exp{(\sigma_2^2)} - 1) \quad (41)$$

This equation shows that the coefficient of variation of $Y(t)$ is constant (does not vary with t) and that if σ_2^2 is small, the mean of $Y(t)$ is close to $C(t)$.

Another aspect of the statistical properties of the data not illustrated in the above example is the correlation structure present in the data. It is frequently assumed that the data values are uncorrelated (or conditionally uncorrelated). Whether or not one takes a logarithm of uncorrelated data does not introduce a correlation. Hence, in the example above, there was a difference in mean and variance between the transformed and untransformed models, but no difference in the correlation structure. However, some data transformations introduce changes in the mean, variance, and correlation structure. Deconvolution of the plasma drug concentration data is an example of such a transformation because each deconvoluted value is a function of plasma drug concentrations at all of the earlier times. As a consequence, deconvoluted data would be expected to exhibit a strong correlation structure and assuming that such data were uncorrelated (16,22,28,29) would be difficult to defend. For example, consider a situation in which 10 measurements of plasma drug concentration are made over a 24-hour period. Homoscedastic, uncorrelated measurements of the plasma drug concentrations were simulated and deconvoluted using Wagner–Nelson deconvolution. The 10 deconvoluted values have increasing variance, as illustrated in Figure 8, and are correlated, as shown in Table 3. It was pointed out in the section "The Functional Relationship" that convolution models have deconvolution equivalents. This equivalence is frequently lost in the course of constructing

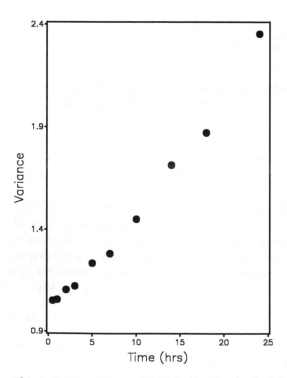

Figure 8 The variance of the Wagner–Nelson deconvoluted values at 10 sampling times over 24 hours. These values are based on simulated plasma drug concentration measurements at the same 10 sampling times, which were uncorrelated and homoscedastic. This figure demonstrates that the deconvoluted values are not homoscedastic.

Table 3 The Pearson Correlation Matrix for the Wagner–Nelson Deconvoluted Data

1.00	0.05	0.04	0.05	0.04	0.05	0.04	0.04	0.04	0.04
0.05	1.00	0.07	0.08	0.07	0.07	0.07	0.06	0.06	0.05
0.04	0.07	1.00	0.11	0.09	0.09	0.09	0.08	0.08	0.07
0.05	0.08	0.11	1.00	0.14	0.14	0.14	0.13	0.12	0.11
0.04	0.07	0.09	0.14	1.00	0.20	0.19	0.17	0.16	0.15
0.05	0.07	0.09	0.14	0.20	1.00	0.26	0.24	0.23	0.20
0.04	0.07	0.09	0.14	0.19	0.26	1.00	0.34	0.33	0.29
0.04	0.06	0.08	0.13	0.17	0.24	0.34	1.00	0.42	0.37
0.04	0.06	0.08	0.12	0.16	0.23	0.33	0.42	1.00	0.49
0.04	0.05	0.07	0.11	0.15	0.20	0.29	0.37	0.49	1.00

Note: The simulated plasma drug concentration data were not correlated with each other, and this table shows that the corresponding deconvoluted data are correlated. Each column and each row corresponds with one of the 10 sampling times (time increasing from left to right and from top to bottom). For example, the value 0.42 in the cell corresponding with the intersection of the 8th row and 9th column (and vice versa) is the correlation between the Wagner–Nelson deconvoluted values at the 8th and 9th sampling times. As expected, the correlation increases as we move from earlier time points toward later times. The diagonal elements are all unity because each variable is perfectly correlated with itself.

the structural model. Consider the convolution model in Equation 8 and the Wagner–Nelson deconvolution model in Equation 10. These functional models are equivalent. Suppose the structural model based on Equation 8 assumes that the observed plasma drug concentrations are uncorrelated and the structural model based on the Wagner–Nelson deconvolution model assumes that the deconvoluted plasma drug concentration data are uncorrelated, then the equivalence between the two functional models is not preserved in the structural models.

Another transformation that is frequently used in IVIVC analysis is data averaging. Both in vitro and in vivo data are commonly averaged across dosage units and subjects at each sampling time point (6,16,17,26). This transformation results in a loss of information because we lose the ability to distinguish between dosage units and between subjects and, as a result, cannot quantify the variation between them. In addition, a curve based on averages and the individual curves whose values were averaged in order to construct it could have very different shapes. This is illustrated in Figure 9 using simulated dissolution/time curves. In addition to this, there is also an inconsistency in averaging the data before either convoluting, deconvoluting it, or applying a differential equation to it. This is because as far as the in vivo data are concerned the LTID system of interest is an individual subject and, consequently, the convolution, deconvolution, or the differential-equation-based model should be applied to individual subjects rather than to averaged data.

As mentioned previously, the actual fraction of the dosage unit that has dissolved in vitro or in vivo must take values between 0 and 1. However, the measured fraction dissolved in vitro may (and frequently does) take values

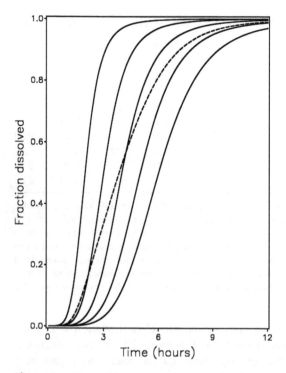

Figure 9 The *solid lines* represent simulated dissolution curves for individual dosage units. The *dashed line* is the curve joining the mean values of the fraction dissolved at each time point. Note that the shape of the mean curve is different from the curves for the individual dosage units.

greater than one. Part of the explanation for this is that there are errors associated with the process of sampling the dissolution medium and the subsequent drug assay. Another reason may be that the measured fractions are fractions of a constant value (e.g., label claim) rather than fractions of the dose actually contained in the dosage unit. Whilst the functional model needs to take account of the fact that the actual fraction lies between 0 and 1, the structural model may need to accommodate measured fractions greater than unity.

The fact that both the in vitro and in vivo data are "repeated measures" data (33) is an important consideration in constructing the structural model. This is because the data consist of measurements made repeatedly across time on the same experimental unit (dosage unit or subject). Such data are undoubtedly correlated. Consider, for example, a subject who is a good (or bad) absorber of the drug. Then the subject's plasma drug concentrations will tend to be consistently high (or low), and are therefore positively correlated with one another. The structural model should take account of this correlation. One way in which this can be done is using random effects models (7), but there are other options available (10,33).

PARAMETER ESTIMATION

Once the data have been collected and the structural model has been constructed, the next step is to use the data to estimate the parameters (unknown constants) in the model. This process is often described as "model fitting" because it amounts to finding the parameter estimates, which give the best (in some sense) agreement between the model and the data. There are a number of alternative approaches available, including maximum likelihood (34), least squares (35) and Bayesian analysis (36). Irrespective of which method is used it is based on the structural model and, consequently, different structural models give rise to different parameter estimates. For example, ordinary (unweighted) least squares is appropriate for homoscedastic uncorrelated data, whereas generalized or weighted least squares (37) is appropriate for correlated heteroscedastic data. Space limitations preclude any detailed discussion of these parameter estimation methods, the interested reader should consult the above references for details.

DISCUSSION AND CONCLUSIONS

Some examples of functional models were described in the section "The Functional Relationship" where the convolution, deconvolution, and differential-equation-based models are equivalent. The deconvolution and convolution functional models should always be equivalent since one is essentially the "inverse" of the other. However, the convolution and differential-equation-based models can have some important differences. First, the convolution model assumes that the system is linear, whereas differential equations can be used to describe nonlinear systems. Consider, for example, a one-compartment model with saturable drug elimination. In such a case, the system is nonlinear and the convolution integral should not be used to describe it. However, the model based on the appropriate differential equations is still applicable. On the other hand, the differential equation(s) require the complete specification of the structure of the system, whereas the convolution integral can be used with the saturated model for the unit impulse response that does not specify any of the system's structure. In addition, the differential equations often require multiple numerical integration steps, whereas the convolution integral has just one integration to be performed. The degree of model mis-specification involved when the convolution integral is used to describe a nonlinear system depends on the extent of the nonlinearity. For many nonlinear systems, the convolution integral might be a reasonable approximation.

Having said that convolution and deconvolution functional models are equivalent, it must be stressed that the corresponding structural models are rarely, if ever, equivalent. This is due to the fact that they make very different assumptions about the statistical properties of the data. Deconvolution-based methods frequently assume that the deconvoluted data are homoscedastic and uncorrelated. Given that each deconvoluted value is a function of the plasma drug concentration data at all earlier times, these assumptions are unrealistic as demonstrated by the example in Figure 8 and Table 3.

The transition from functional to structural model is based on assumptions about the statistical properties of the data. This is a step in the model building process, which is frequently not given the attention it deserves or is ignored completely. What must be borne in mind is that model building aims to construct a model, that describes the data that has been collected and must therefore address the statistical properties of the data. The functional model, which describes relationships between underlying variables, is of course central to the model, but the fact that the data collected are only estimates of these variables should be borne in mind. In other words, the (random) differences between the data values and the underlying quantities must be included in the model by specifying their statistical properties (7,10,30).

The in vivo part of all the models discussed earlier is based on a systems approach where the system is an individual subject or patient. Clearly then, these models ought to be applied to individual subject data rather than to data averaged across subjects, as is often the case.

Another approach to IVIVC analysis that has not been discussed in this chapter is that based on artificial neural networks (ANN) (38,39,40). This is a very different approach to any of those that have been discussed earlier. All of the models described earlier could be labeled as semi-mechanistic because they are at least partly based on an assumed mechanism for the processes underlying the data. The ANN approach is completely empirical in nature and therefore has little or nothing in common with the models and methods discussed in this chapter.

ACKNOWLEDGMENTS

The author wishes to acknowledge the very helpful discussions with John Davis, Clare Gaynor, Siobhán Hayes, and Theresa Shepard.

REFERENCES

1. Kendall M, Stuart A. The Advanced Theory of Statistics. Vol. 2. 4th ed. London: Griffin & Co, 1979:399–440.
2. Finkelstein L, Carson ER. Mathematical Modelling of Dynamic Biological Systems. Oregon: Research Studies Press, 1979.
3. Food and Drug Administration. Guidance for industry: extended release oral dosage forms: development, evaluation, and application of in vitro/in vivo correlations 1997.
4. European Agency for the Evaluation of Medicinal Products. Note for guidance on quality of modified release products: A: oral dosage forms B: transdermal dosage forms 2000.
5. Balan G, Timmins P, Greene DS, Marathe PH. In vitro-in vivo correlation (IVIVC) models for metformin after administration of modified-release (MR) oral dosage forms to healthy human volunteers. J Pharm Sci 2001; 90:1176–1185.

6. Sirisuth N, Augsburger LL, Eddington ND. Development and validation of a non-linear IVIVC model for a diltiazem extended release formulation. Biopharm Drug Dispos 2002; 23:1–8.
7. O'Hara T, Hayes S, Davis J, Devane J, Smart T, Dunne A. In vivo-in vitro correlation (IVIVC) modeling incorporating a convolution step. J Pharmacokinet Pharmacodyn 2001; 28:277–298.
8. Gillespie WR. Convolution-based approaches for in vivo-in vitro correlation modeling. In: Young D, Devane JG, Butler J, eds. Advances in Experimental Medicine and Biology. Vol. 423. In vitro-In vivo Correlations. New York: Plenum Press, 1997:53–65.
9. Liu Y, Schwartz JB, Schnaare RL, Sugita ET. A multi-mechanistic drug release approach in a bead dosage form and in vitro/in vivo correlations. Pharm Dev Technol 2003; 8:409–417.
10. Mauger DT, Chinchilli VM. In vitro-in vivo relationships for oral extended-release drug products. J Biopharm Stat 1997; 7:565–578.
11. Veng-Pedersen P, Gobburu JVS, Meyer MC, Straughn AB. Carbamazepine level A in vivo-in vitro correlation (IVIVC): A scaled convolution based predictive approach. Biopharm Drug Dispos 2000; 21:1–6.
12. Aiache JM, Islasse M, Beyssac E, Aiache S, Renoux R, Kantelip JP. Kinetics of indomethacin release from suppositories. In vitro-in vivo correlation. Int J Pharm 1987; 39:235–242.
13. Buchwald P. Direct, differential-equation-based in vitro-in vivo correlation (IVIVC) method. J Pharm Pharmacol 2003; 55:495–504.
14. Wagner JG, Nelson E. Kinetic analysis of blood levels and urinary excretion in the absorptive phase after single doses of drug. J Pharm Sci 1964; 53:1392–1394.
15. Loo JCK, Riegelman S. New method for calculating the intrinsic absorption rate of drugs. J Pharm Sci 1968; 57:918–928.
16. Eddington ND, Marroum P, Uppoor R, Hussain A, Augsburger L. Development and internal validation of an in vitro-in vivo correlation for a hydrophilic metoprolol tartrate extended release tablet formulation. Pharm Res 1998; 15:466–473.
17. Polli JE, Crison JR, Amidon GL. Novel approach to the analysis of in vitro-in vivo relationships. J Pharm Sci 1996; 85:753–760.
18. Verotta D. A general framework for non-parametric subject-specific and population deconvolution methods for in vivo-in vitro correlation. In: Young D, Devane JG, Butler J, eds. Advances in Experimental Medicine and Biology. Vol. 423. In vitro-In vivo Correlations. New York: Plenum Press, 1997:43–52.
19. Caramella C, Ferrari F, Bonferroni MC, et al. In vitro/in vivo correlation of prolonged release dosage forms containing diltiazem HCl. Biopharm Drug Dispos 1993; 14:143–160.
20. Rohrs B, Skoug JW, Halstead GW. Dissolution assay development for in vitro-in vivo correlations. In: Young D, Devane JG, Butler J, eds. Advances in Experimental Medicine and Biology. Vol. 423. In vitro-In vivo Correlations. New York: Plenum Press, 1997:17–30.
21. Sathe P, Tsong Y, Shah V. In vitro dissolution profile comparison and IVIVR. In: Young D, Devane JG, Butler J, eds. Advances in Experimental Medicine and Biology. Vol. 423. In vitro-In vivo Correlations. New York: Plenum Press, 1997:31–42.
22. Kortejarvi H, Mikkola J, Backman M, Antila S, Marvola M. Development of level A, B and C in vitro-in vivo correlations for modified-release levosimendan capsules. Int J Pharm 2002; 241:87–95.

23. Langenbucher F. Handling of computational in vitro/in vivo correlation problems by Microsoft Excel II. Distribution functions and moments. Eur J Pharm Biopharm 2003; 55:77–84.

24. Rostami-Hodjegan A, Shiran MR, Tucker GT, et al. A new rapidly absorbed paracetamol tablet containing sodium bicarbonate. II. Dissolution studies and in vitro/in vivo correlation. Drug Develop Ind Pharm 2002; 28:533–543.

25. Ishii K, Saito Y, Itai S, Takayama K, Nagai T. In vitro dissolution test corresponding to in vivo dissolution of sofalcone formulations. S.T.P. Pharma Sci 1997; 7:270–276.

26. Uppoor VRS. Regulatory perspectives on in vitro (dissolution)/in vivo (bioavailability) correlations. J Contr Release 2001; 72:127–132.

27. Abuzarur-Aloul R, Gjellan K, Sjolund M, Graffner C. Critical dissolution tests of oral systems based on statistically designed experiments. II. In vitro optimization of screened variables on ER-coated spheres for the establishment of an in vitro/in vivo correlation. Drug Develop Ind Pharm 1998; 24:203–212.

28. Mendell-Harary J, Dowell J, Bigora S, et al. Nonlinear In vitro-In vivo Correlations. In: Young D, Devane JG, Butler J, eds. Advances in Experimental Medicine and Biology, Vol. 423. In vitro-In vivo Correlations. New York: Plenum Press, 1997:199–206.

29. Bigora S, Piscitelli D, Dowell J, et al. Use of nonlinear mixed effects modelling in the development of in vitro-in vivo correlations. In: Young D, Devane JG, Butler J, eds. Advances in Experimental Medicine and Biology, Vol. 423. In vitro-In vivo Correlations. New York: Plenum Press, 1997:207–215.

30. Dunne A, O'Hara T, Devane J. Approaches to IVIVR modelling and statistical analysis. In: Young D, Devane JG, Butler J, eds. Advances in Experimental Medicine and Biology. Vol. 423. In vitro-In vivo Correlations. New York: Plenum Press, 1997a:67–86.

31. Dunne A, O'Hara T, Devane J. Level A in vivo-in vitro correlation: nonlinear models and statistical methodology. J Pharm Sci 1997b; 86:1245–1249.

32. Johnson NL, Kotz S. Distributions in Statistics. New York: John Wiley, 1970: 112–132.

33. Diggle PJ, Liang KY, Zeger SL. Analysis of Longitudinal Data. Oxford: Oxford University Press, 1996.

34. Pawitan Y. In all likelihood: statistical modelling and inference using likelihood. Oxford: Oxford University Press, 2001.

35. Bates DM, Watts DG. Nonlinear regression analysis and its applications. New York: John Wiley & Sons, 1988.

36. Berger JO. Statistical decision theory and Bayesian analysis. New York: Springer-Verlag, 1985.

37. Montgomery DC, Peck E. Introduction to Linear Regression Analysis. New York: John Wiley & Sons 1992:378–381.

38. Hussain AS. Artificial neural network based in vitro-in vivo correlations. In: Young D, Devane JG, Butler J, eds. Advances in Experimental Medicine and Biology. Vol. 423. In vitro-In vivo Correlations. New York: Plenum Press, 1997:149–158.

39. Dowell JA, Hussain AS, Stark P, Devane J, Young D. Development of in vitro-in vivo correlations using various artificial neural network configurations. In: Young D, Devane JG, Butler J, eds. Advances in Experimental Medicine and Biology. Vol. 423. In vitro-In vivo Correlations. New York: Plenum Press, 1997:225–239.

40. Dowell JA, Hussain AS, Devane J, Young D. Artificial neural networks applied to the in vitro-in vivo correlation of an extended-release formulation: initial trials and experience. J Pharm Sci 1999; 88:154–160.

4

The Role of In Vitro–In Vivo Correlation in Product Development and Life Cycle Management

Shoufeng Li

Pharmaceutical and Analytical Development, Novartis Pharmaceuticals Corporation, East Hanover, New Jersey, U.S.A.

Martin Mueller-Zsigmondy

Pharmaceutical and Analytical Development, Novartis Pharmaceuticals AG, Basel, Switzerland

Hequn Yin

Early Clinical Development, Novartis Pharmaceuticals Corporation, East Hanover, New Jersey, U.S.A.

OVERVIEW: IN VITRO–IN VIVO CORRELATION—A FOUR-TIER APPROACH IN PRODUCT DEVELOPMENT

The development of a new chemical entity (NCE) usually undergoes various stages (Fig. 1). Our knowledge of the physicochemical and biopharmaceutical properties of the molecule generally improves as it progresses through the development stages. Although many in vivo tests are carried out in drug discovery stage, main focus at this stage is the efficacy of the molecule rather than its development potential. Due to large number of molecules and limited physicochemical information, in-silico simulation based on structure or high-throughput experimental data is often used. However, the developability concept has become ever more important over the last decade, while biopharmaceutical properties

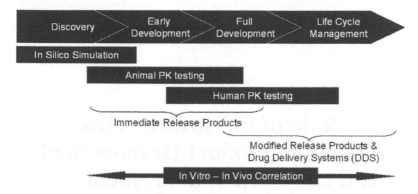

Figure 1 Discovery and development phases of new chemical entity and application of in vitro–in vivo correlation in drug development process. *Abbreviation*: PK, pharmacokinetic.

are among the most important components (1,2). From preclinical development until proof of concept (PoC) initiation, pharmacokinetic (PK) testing is often carried out. PK of different physical forms, salts, and particle sizes of drug molecule can be evaluated in a preclinical animal model. This provides the first opportunity to correlate the in vitro measurement (i.e., dissolution of the molecule) to its in vivo performance, such as C_{max}, area under the concentration time curve (AUC) or deconvoluted, in vivo dissolution profiles. In vivo animal PK data of different physical forms, salts, and particle sizes or formulations also provide the first opportunity for the development of a biorelevant dissolution method. In a recent review published by Li et al. (3), a decision tree for dissolution testing design based on biopharmaceutics classification system (BCS) and physicochemical properties of the molecule has been proposed. This provides guidance on dissolution method setting, which can be further validated in full development once human PK data are available. Since formulation development and optimization at a later stage rely on the dissolution method established early on, it is essential the selection of the dissolution method is as meaningful and relevant as possible. When clinical data on different formulations or different particle sizes of the drug substance are available, additional investigations can be performed to verify whether the dissolution test method need be modified or challenged. There is a tremendous scientific and practical value when an in vitro–in vivo correlation (IVIVC) can be established using human PK data, but this often involves a cross-functional team of scientists from formulation, dissolution, and clinical PK development. A significant amount of information is already available in the literature, and successful IVIVCs have been demonstrated for modified release (MR) formulations, on the basis of the 1997 Food and Drug Administration guidelines in which the procedure and acceptance criteria for a successful IVIVC has been clearly defined (4). In the life-cycle management (LCM) stage of drug development, IVIVC is even more important

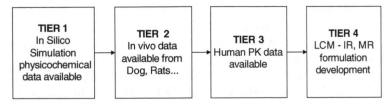

Figure 2 Four-tier approach for in vitro–in vivo correlation development. *Abbreviations*: PK, pharmacokinetic; LCM, life cycle management; IR, immediate release; MR, modified release.

and has to be considered as part of the development strategy for MR or alternative delivery systems including parental depot, transdermal patch, and so on. Several such successful cases have been reported by Young (5) in this area.

In the authors' opinion, application of IVIVC in new drug development process can be achieved in a four-tier approach as illustrated in Figure 2. In silico simulation based on compound structure and limited physicochemical information can be performed in drug discovery stage to rank order the absorption potential of a molecule (Tier 1), followed by correlating in vitro dissolution with its in vivo performance in preclinical setting (Tier 2), the dissolution method developed at this stage can be further validated using human PK data (Tier 3), and finally, human PK data from various sources can be systematically utilized to establish a valid IVIVC model to further support future formulation development during life cycle management (Tier 4). The four-tier approach is proposed based on stages of NCE product development. As discussed earlier, IVIVC development is an evolving process, which should be perfected through product development and it is important that such a concept be applied to product development as early as possible.

INTRODUCTION

Categories of In Vitro–In Vivo Correlation

Levels A, B, and C IVIVCs are clearly defined in regulatory guidances (4,6) with the Level A correlation, as point-to-point correlation, which considers complete in vivo and in vitro profiles being the preferred correlation. In the Level A correlation, there is no omission of information from either in vivo or in vitro data. A convoluted plasma concentration profile can be calculated from an in vitro dissolution profile. The benefit of such a correlation includes, amongst other things, the possibility to replace a bioequivalence study with comparative in vitro dissolution data (7). To develop a valid Level A correlation that can be accepted by the agency for so-called biowaivers, an IVIVC model needs to be developed for at least two formulations, with three or more formulations being preferred. The formulations should have significant differences in their in vitro and in vivo behavior. Validation (i.e., internal and external predictability of the model) also needs to be demonstrated (8).

The United States Pharmacopeia (USP) also refers to the above-mentioned categories, but does not mention internal and external predictability (9). Some references describe Level A correlation that consider only one formulation, which do not allow the model to be applied to other formulations or more batches of the same formulations. In this respect, many IVIVC categorized as Level A are not recognized by the regulatory agency (10). Such correlations, nevertheless, help to assist in the dissolution method development and have the potential to be developed into a true Level A correlation which can be recognized by the agency.

In the case of Level B correlations, the entire course of in vitro and in vivo profiles are also considered, with the information in the profiles reduced to a single parameter. Level B correlation is carried out where the mean in vitro dissolution time is compared with the mean in vivo dissolution time or mean in vivo residence time (11,12). However, since the entire plasma concentration profile cannot be predicted based on in vitro dissolution data, the benefit of Level B correlation is therefore limited and is not accepted by authorities for biowaivers.

In the last category, Level C correlation, only one point is taken from the profiles. A typical example of a Level C correlation is the correlation of a percentage of drug released at a certain time point with C_{max}. Immediate release (IR) formulations of good water-soluble substances are usually characterized by this manner (13,14). For MR formulations and IR products with less water-soluble drug substance, correlation of partial AUC with drug release at certain time point is rational, since in this case, the dissolution process is usually longer than the gastrointestinal (GI) transit time of the pharmaceutical form, and AUC can be reduced because the drug is not being fully released (15–17). In early formulation development stage, a Level C correlation can provide valuable information. In the later phase of development and after process transfer to production, its usefulness becomes limited because the entire plasma concentration profile cannot be predicted, unless a multiple Level C correlation can be established, under which scenario a Level A correlation also becomes very likely.

DEVELOPMENT OF IN VITRO–IN VIVO CORRELATIONS

Deconvolution

In the early days of IVIVC, it was suggested that a correlation should be implemented only for comparable data (18), where comparison was made between actual data measured in vitro, simulated in vivo data, and vice versa. Correlations are obtained when the in vivo plasma concentration profile is converted via mathematical modeling using model-dependent or model-independent deconvolution into the in vivo absorption profile. The in vivo absorption profile, which is often identical to the in vivo dissolution profile, is then used to establish an IVIVC.

Due to the intrinsic difference between dissolution conditions in vitro and in vivo, the in vitro dissolution profiles can be scaled by mathematical means, represented by Equation 1 (19):

$$X_{\text{vivo}}(t) = a_1 + a_2 \times X_{\text{vitro}}(b_1 + b_2 \times t) \quad \text{if } t > T \text{ then } t = T$$
$$\text{if } b_1 < b_2 \times t \text{ then } b_1 + b_2 \times t = 0$$

(1)

whereby $X_{\text{vivo}}(t)$ represents the absorption profile as a function of time and $X_{\text{vitro}}(t)$ represents the dissolution profile. The modifications to the in vitro profile as a function of time t is achieved by introduction of a time scale factor b_2, if the in vitro dissolution process occurs faster or slower than the corresponding in vivo dissolution; or through a lag time b_1 to allow an initial lag time in the in vivo absorption because of necessary preabsorption transit through the stomach; and of a cut-off factor T to accommodate dissolution slower than the GI transit times. The actual correlation is obtained via comparison of the in vivo profile with the scaled in vitro dissolution profile by a linear regression, which provides the slope a_2 and the intercept a_1 as a link function between both profiles.

The aforementioned calculation from the in vivo profile to the in vivo absorption/dissolution profile is known as deconvolution (output to input) and is illustrated in Figure 3. The classical methods of deconvolution of plasma profiles include Wagner–Nelson (10,20–22), Loo–Riegelman (23,24) and numerical deconvolution (8,25,26). The Wagner–Nelson method is a model-dependent method based on one-compartment model, it has a great advantage of not requiring additional in vivo data except oral plasma profile. The Loo–Riegelman method is based on two-compartment model, which requires intravenous dosing data. Model-independent numerical deconvolution requires in vivo

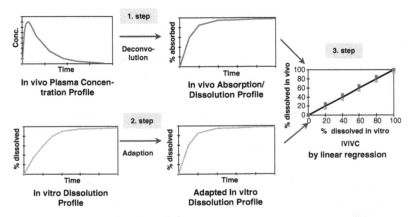

Figure 3 The classical three steps of deconvolution. *Abbreviation:* IVIVC, in vitro–in vivo correlation.

plasma data from an oral solution or intravenous as impulse function for the application. All three methods have their limitations, but the requirement of additional data in addition to oral plasma data from a tablet or capsule significantly limit the application of the later two methods. There are numerous literature examples that use model-independent methods (27,28), Wagner–Nelson, or Loo–Riegelman methods (29). Convolution and deconvolution by means of excel sheets are also described by Langenbucher (23,30,31).

Convolution

Conversion of the in vitro dissolution profile to a plasma concentration profile can take place via convolution (input to output). Recently, convolution methods have been established, which convolutes the in vitro dissolution profiles without implementing the correlation of the in vivo absorption/dissolution profile with the in vitro dissolution profile (i.e., physiology based model and simulation software). The model uses multiple differential equations representing various physiological events (32) and convolution-based methods (33,34). Such a convolution does not take place in several partial stages, rather in one single stage that is schematically illustrated in Figure 4.

A major advantage of convolution-based methods for IVIVC is that no additional in vivo data such as intravenous injections or oral solutions are required. However, these methods can only mathematically fit the data by minimizing the squared error; even though the results obtained are mathematically correct it may not be meaningful PK or physiological models. A critical

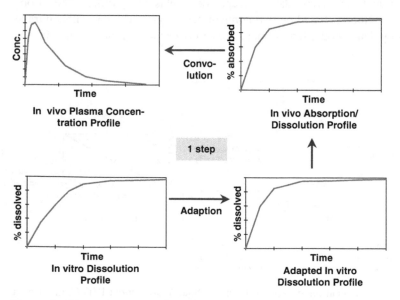

Figure 4 The one-step procedure of a convolution.

assessment of the calculated parameters is absolutely necessary. Further, the fitting procedure should be performed several times with different starting values, in order to avoid reaching a local minimum. Last but not the least, these methods should be optimized to as few variables as possible, as the fitting procedure becomes more complex and error-prone with more variables.

Basic Principles of In Vitro–In Vivo Correlation

The in vitro dissolution and in vivo absorption/dissolution profiles play a key role in the development of an IVIVC. Characterization of these profiles warrants a detailed discussion. The cumulative dissolution curves can be well represented by parameters that describe extent of dissolution, time delays, and shape of the profile. The Weibull function depicted in Equation 2 is one of the models that is suitable for such purpose (21,35,36):

$$F(t) = F^{\infty} \cdot \left(1 - e^{((t+t_0)/\alpha)^{\beta}}\right) \tag{2}$$

where α represents the time at which 63.2% of the drug is dissolved, β is a shape factor that, at values below 1, yields a curve with an initially steep slope followed by a flat course; at a value equal to 1, it describes an exponential curve; and at values greater than 1, yields a curve with a sigmoidal shape. Various shape factors can also be interpreted as different release mechanisms. F^{∞} is the dissolved fraction of the dose after an infinite time. t_0 is lag time that considers the delayed start of dissolution process. A perfect correlation can be achieved if all parameters of the Weibull function of in vivo and in vitro profiles are identical.

For example, an in vitro dissolution profile has the following characteristics: $F^{\infty} = 100$, $t_0 = 0$, $\alpha = 1$, and $\beta = 0.5$; whereas the same formulation, when tested in vivo, its in vivo dissolution profile is characterized by the following Weibull parameters: $F^{\infty} = 100$, $t_0 = 0$, $\alpha = 1$, and $\beta = 1.5$ (Fig. 5).

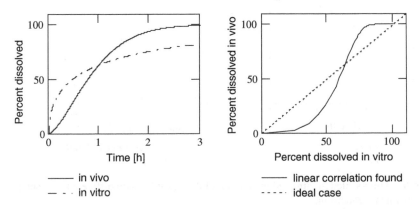

Figure 5 In vitro and in vivo dissolution profiles with different shapes and their linear regression.

F^{∞}, t_0, and α have identical values, but β (shape factor) is distinctly different. When percentage dissolved in vivo is plotted against percentage dissolved in vitro, a nonlinear relationship deviating greatly from ideal linear curve is obtained (Fig. 5). An IVIVC with linear correlation cannot be established.

Whenever appropriate, polynomial functions can be used to obtain non-linear IVIVC. If there are factors other than in vivo dissolution contributing to the absorption, the usefulness of the IVIVC obtained by nonlinear regression can be very limited. The validity of the correlation has to be verified using the internal and external prediction errors. One could also choose to modify the in vitro dissolution test condition to obtain improved, that is, linear IVIVC.

Similarity between in vitro and in vivo profiles in addition to time factor α is necessary for a successful IVIVC. In the example given earlier, the in vivo and in vitro profiles have the same time factor α. In most cases, however, the in vitro profile may be faster or slower than the in vivo profile. For instance, an in vivo profile ($F^{\infty} = 100$, $t_0 = 0$, $\alpha = 1$, and $\beta = 1.5$) is compared with an in vitro profile ($F^{\infty} = 100$, $t_0 = 0$, $\alpha = 0.2$, and $\beta = 1.5$). F^{∞}, t_0, and β have identical values, but the time factor α is distinctly different. When the percentage dissolved in vivo is plotted versus the percentage dissolved in vitro, a nonlinear correlation as shown in Figure 6 is obtained. When a time scale factor b_2 is used, which stretches the x-axis by a factor of 5, the in vitro profile can be scaled and fitted very well with the in vivo profile. Equation 3, which is a simplification of Equation 1 with $b_1 = 0$, $a_2 = 1$, and $a_1 = 0$, describes the mathematical relationship with $b_2 = 0.2$:

$$X_{\text{vivo}}(t) = X_{\text{vitro}}(b_2 \times t) \tag{3}$$

Using the time-scaled in vitro data, a perfect linear correlation, as shown in Figure 7, can be obtained.

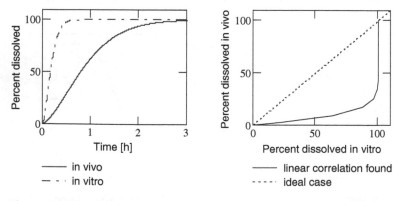

Figure 6   Time-scaled in vitro and in vivo dissolution profiles with different time factors and their linear regression.

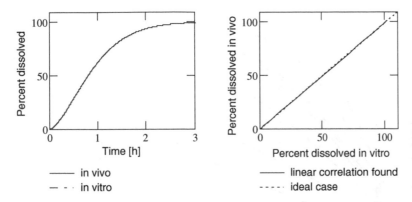

Figure 7 Time-scaled in vitro and in vivo dissolution profiles with different time factors and their linear regression.

The introduction of a time scale factor is acceptable as long as it is used for all formulations and for all further applications of the IVIVC model. The time scale factor can be determined by plotting the time needed for in vivo dissolution versus the time needed for in vitro dissolution of a particular amount of drug from the dosage form (Levy plot). After linear regression forced through zero, the reciprocal of the slope of the regression line is the time factor one should use for IVIVC development. Examples for the calculation and application of a time factor are given in the literature and in this book (37,38).

Alternatively, the time scale factor can be calculated as the ratio of the time factors α of in vitro and in vivo Weibull fitted profiles, provided both profiles have an infinite dissolution F^∞ of 100%.

$$b_2 = \frac{\alpha_{\text{in vitro}}}{\alpha_{\text{in vivo}}} \tag{4}$$

Application of the time scale factor can be demonstrated by the following example. Various dissolution conditions were evaluated for an in vivo relevant method (Fig. 8), where pH, the agitation speed, and ionic strength of the dissolution medium were varied. The dependence of dissolution on these variables is clearly demonstrated. The Weibull parameters of these profiles can be determined and listed in Table 1. The similar β values of all in vitro profiles with distinctly different time factors is evident. If a time scaling, based on Equation 4, with the factors given in Table 1 is implemented for each individual in vitro profile, the time-scaled profiles shown in Figure 9 are obtained, where the profiles are almost matching with each other. All profiles obtained under these conditions are suitable in order to achieve an IVIVC, except time scale factors used are different.

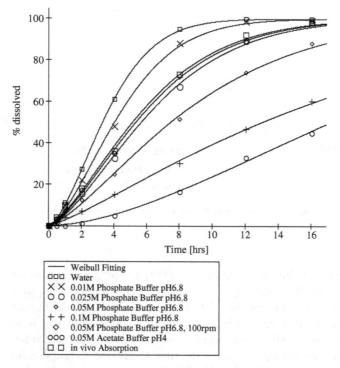

Figure 8 Dissolution profiles depending on pH value, agitation speed, and ionic strength of the medium.

Another example in determining the time factor using a Levy plot is shown in Figure 10, where the line deviates upwards after a specific time. Such phenomenon can be explained by the permeability or dissolution change on transit of formulation from the stomach to the intestine and into the colon (39).

Table 1 Weibull Parameters and Time Scale Factors

Operating conditions	α	β	Time scale factor b_2
Water, 50 rpm	4.11	1.58	0.62
0.01M Phosphate buffer pH 6.8, 50 rpm	5.06	1.54	0.77
0.025M Phosphate buffer pH 6.8, 50 rpm	7.23	1.47	1.10
0.05M Phosphate buffer pH 6.8, 50 rpm	9.73	1.37	1.47
0.1M Phosphate buffer pH 6.8, 50 rpm	17.39	1.25	2.64
0.05M Phosphate buffer pH 6.8, 100 rpm	6.83	1.44	1.04
0.05M Acetate buffer pH 4, 50 rpm	21.21	1.72	3.22
In vivo absorption	6.60	1.42	—

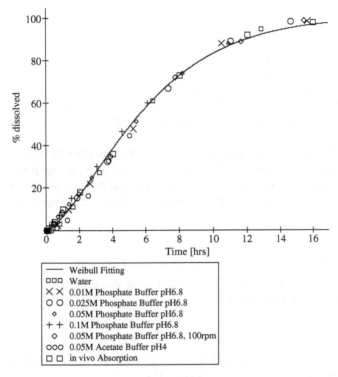

Figure 9   Time-scaled dissolution profiles.

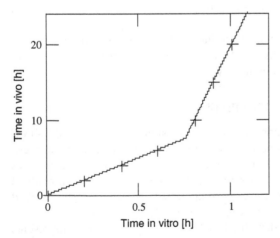

Figure 10   Levy plot with upward curvature indicating change in release mechanism in gastrointestinal tract.

Cut-Off Factor and Function

If the in vivo dissolution is too slow compared to GI transit time, it can result in reduced absorption (40). If there are regional absorption windows, only partial absorption is possible, despite complete in vivo dissolution. In such cases, the in vivo dissolution profile is not directly related to the in vivo absorption profile; therefore, the in vivo dissolution profile must be further adapted to match the absorption profile, which can be achieved by introducing a cut-off factor.

The cut-off factor T can be included as an if-then condition in the IVIVC, as a modification of Equation 5. Alternatively, incorporation of a truncating function in Equation 5 is also possible, which causes a rather gentler subsidence in absorption and can also be used for the numerical solution of differential equations.

$$\varphi(x) = \frac{e^{-\tau(t-T)}}{1 + e^{-\tau(t-T)}} \tag{5}$$

where $\varphi(x)$ is the truncating function, τ is a measure for the steepness of the function, and T is the cut-off time point.

Lag Time

Introduction of a lag time t_0 in Equation 1 is required when there is a delay of in vivo release compared with the in vitro release (41).

For initial data points, when t is smaller than t_0, the resulting value can be negative. This can be overcome by an if-then condition as shown in Equation 1, or bypassed by the following expression in Equation 6:

$$\text{pos}(x) = \frac{1}{2} \times (x + |x|) \quad \text{with } \text{pos}(x + t_0) \tag{6}$$

This form of expression can be helpful for lag time, particularly with differential equation based IVIVCs, when the mathematical program used does not accept negative expressions or if-then conditions.

Correlation of In Vitro and In Vivo Profiles

Correlation of profiles by means of linear regression is the classical IVIVC method. Altering in vitro test conditions systematically by statistical experimental design is also a very effective tool to match the dissolution characteristics of the in vivo dissolution of the formulations (41–43). This approach enabled Qiu et al. (44) to achieve a good linear correlation between percentage absorbed and percentage dissolved of three controlled release formulations.

An alternative method is described by Polli et al. (45), where it represents an extension of linear correlation. For IR formulations in which absorption may be partially permeability limited or regional dependent, a nonlinear correlation may provide certain advantages (36). Further correlation of in vivo dissolved dosages with in vitro dissolved dosages is described by Dunne et al. (46) by the use of odds, hazard, or reversed hazards functions.

THE ROLE OF IN SILICO SIMULATION IN EARLY DEVELOPMENT

In the early research and development stages, one of the frequently asked questions is "What is the oral absorbability of a molecule?" Medicinal chemists who synthesize potential pharmaceutical structures like to know the likelihood of oral absorption even before a molecule is synthesized. Formulation experts, when approached by others to enhance absorption, like to ask what the extent of oral absorption is in order to assess if formulation is really the problem and something they should work on. However, proper in vivo assessment of oral absorption usually requires a radio-labeled molecule, is relatively labor intensive and expensive, and usually takes place much later in the research and development scheme. Additionally, such in vivo experiments involving both intravenous and oral administration are typically done in animal species and are rarely performed in humans for orally administered drugs, for the purpose of assessing the oral bioavailability of a molecule. Therefore, it is desirable to explore in silico approaches to allow for an earlier assessment of absorbability, to extrapolate to situations where experimental data are not available, and to help identify and focus on limiting factors of absorption.

Currently there are two in silico approaches for the predicting absorption: statistical models and mechanism-based models (3). Statistical models are based on a statistical relationship between inputs, typically molecular descriptors derived from a molecular structure, and outputs, in this case oral absorption percentages. Mechanism-based models rely on a good understanding of absorption processes including physiology, GI dissolution, transit, and permeation.

In Silico Simulation Based on Structure of a Compound

Chemical structure-based statistical models strive to establish a structure-absorption relationship. Lipinski (47) was the first to point out that poor oral absorption is more likely to occur for a molecule when there are more than 10 hydrogen bond acceptors, 5 hydrogen bond donors, and when the molecular weight is greater than 500, and calculated log P is greater than 5. This set of empirical rules was later on frequently referred to as "the rule of 5." More quantitative approaches relate absorption to calculated molecular descriptors and/or parameters such as dynamic surface area (48), topological structural features (49,50), and a composite set of parameters or force fields (51–53).

The method by Bai et al. (51) was incorporated into a software called OraSpotter™ (ZyxBio, Cleveland, Ohio, U.S.A.) and was evaluated by the author and others using sets of proprietary chemical structures and experimental data. This model used 899 compounds in the training set to relate 28 computed structural descriptors to actual human oral absorption values, reported in the literatures. The input and output of the model are chemical structures and absorption fraction in human, respectively. The absorption fraction, zero to one, was divided into six categories in 0.16 increments. With two test sets of unpublished proprietary compounds, a successful prediction of 79% to 86% was achieved when the predicted values fell within ± one class. This level of prediction is reasonable enough to be useful in rank ordering and prioritization even prior to chemical synthesis of candidate molecules. Calculation using OraSpotter™ is fast and may be appropriate for high throughput purposes. Since the human absorption data used in the training set represent situations where the dose is relatively low (solubility is probably a less frequent problem) and the formulation may be improved or optimized, the model projection is perhaps biased towards the highest achievable fraction absorbed. This statistical model does not describe the mechanistic aspects of drug absorption, nor does it account for formulation differences and dose levels. Therefore, it may serve as a first filter in improving the odds of success very early on in the discovery stages. Subsequent experimental measurement on solubility and in vitro permeability may help further screen and rank order drug candidates.

In Silico Simulation Based on Experimentally Measured Physico-chemical and Biopharmaceutical Data

As drug candidates progress through research and development, in vitro physico-chemical and biopharmaceutical data gradually become available, which can be fed into mechanistic absorption models. Mechanistic models simulate and model the GI absorption process based on "Advanced Compartmental Absorption and Transit" model (54) in which the small intestine is divided into seven compartments with equal transit time, and the stomach and the colon are treated each as a separate compartment. Drug release, dissolution, precipitation, absorption, and transit across the compartments are explicitly described in the form of specific integrated or differential mathematical equations. Such complex models, together with human and animal physiological parameters, are built into commercial softwares such as GastroPlus™ (SimulationsPlus, Lancaster, California, U.S.A.). Input to GastroPlus™ includes some or all of the following:

1. oral dose,
2. solubility-pH profile, and human intestinal permeability, which can be estimated from in vitro Caco-2 or artificial membrane assays,
3. species, GI transit time, GI pH, food status,
4. formulation release profile, particle size,
5. PK parameters species, GI transit time, GI pH, food status, and
6. PK parameters such as volume of distribution (V_d), clearance (CL), microscopic kinetic rate constants, and so on.

The current version 4.0 of the software contains default physiological GI parameters for human, rat, mouse, dog, monkey, rabbit, and cat. Nearly all parameters in the software are under users' control and can be modified as necessary.

Typical output of the simulation includes

1. fraction of dose absorbed in each and combined GI compartments,
2. sensitivity of parameters in affecting absorption, and
3. plasma concentration-time profile when PK parameters are provided.

In terms of predicting the total fraction of drug absorbed, a correct classification rate was also reported to be approximately 70%, by Parrot and Lave (55), using a test set of 28 drugs, and eight out 10 correct classification by the author using a test set of 10 proprietary compounds.

The GastroPlus[TM] program is a quite transparent system that allows user to grasp an overall picture of the transit and absorption process, the ability to optimize certain parameters, and to run hypothesis testing on IVIVC-related questions. One could simulate the impact of dose, solubility, particle size, release profile, stomach pH, etc., on the extent and time-course of GI absorption. In one example, a question was asked whether or not the particle size of a drug candidate can be relaxed from a current 35 μm to approximately 100 μm without affecting its oral bioavailability. A simulation was therefore carried out (Fig. 11), which suggested that the extent of absorption is essentially not sensitive to changes in particle size at least in the range between 35 and 400 μm. This facilitated decision making without having to conduct time and labor consuming experiments. This kind of in silico approach was also used by the author and collaborators for formulation prescreening in which in vitro release

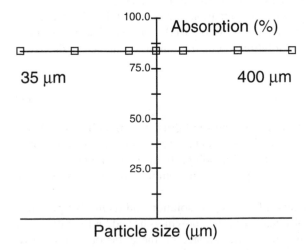

Figure 11 Simulation on the impact of particle size on oral absorption of a drug candidate using GastroPlus[TM].

profile of candidate formulations were used to simulate the absorption outcome and subsequently narrow down the choices prior to in vivo testing. The author and collaborators also had examples of reasonable outcome in using this approach in the design and selection of a particular in vitro release profile to achieve the desired human plasma concentration-time profile. However, more data including in vivo PK parameters were needed in that case.

On the whole, the mechanism-based in silico simulation using physico-chemical and biopharmaceutical data appeared to be educational and practically useful in addressing some of the IVIVC-related questions. This approach could potentially save money and time, and spare resources to focus on more critical issues in pharmaceutical research and development.

IN VITRO–IN VIVO CORRELATION IN PRECLINICAL SETTINGS

As the development phase moves forward, typically during the proof of concept or Phase I stage, the lead candidate is being characterized more thoroughly for its phys-icochemical properties and its ability to develop. A preclinical PK study is usually performed prior to the Phase I clinical trial. A well-designed preclinical PK study, with the input from the formulation and clinical experts, may provide an opportunity to define the scope for drug substance properties, such as particle size, salt forms, and formulations. Combined with simulations, it would also provide an opportunity to identify the rate-limiting factor(s) for absorption. Further, in vivo data obtained at this stage may be used to better justify the dissolution method development. Among the factors that determine the rate and extent of drug absorption following oral administration, dissolution of the solid drug into solution is of primary import-ance in the drug release/absorption process. Factors affecting drug dissolution has been extensively reviewed by Horter (56) as well as Li et al. (3).

Polymorphism, surfactant, complexation, pKa, and GI pH profile are among factors that could influence solubility; while particle size and wetting play major roles in drug dissolution. Drug dissolution can be modeled by Noyes–Whitney equation:

$$\frac{dX}{dt} = \frac{A \times D}{\delta} \times (C_s - X_d/V) \tag{7}$$

where A is the effective surface area of the solid drug, D is the diffusion coeffi-cient of the drug, δ is the effective diffusion boundary layer thickness adjacent to the dissolving surface, C_s is the saturation solubility of the drug under lumenal conditions, X_d is the amount of drug already in solution, and V is the volume of the dissolution medium.

The most common factors influencing dissolution and its in vivo perform-ance include particle sizes, physical form (i.e., polymorphs and/or hydrates), salt forms, and different formulations. The impact of these factors on in vivo performance of the drug product and several case studies will be discussed in the following sections.

In Vivo Performance of Pharmaceutical Salts

A common approach to improve dissolution of a compound is by forming salts. The dissolution of a pH-dependent drug is usually a function of both bulk pH and the surface pH of the solids. Solubility and pKa of the compound often determines its surface pH (57–59).

The processes for systematic screening and selection of the optimal salts with desirable physicochemical and biopharmaceutical properties have been hindered by lack of prediction of their in vivo behavior. Morris et al. (60), Anderson and Flora (61), and Gould (62) have published extensively in this field, intending to provide guidance in selecting an optimal salt form from chemical and biopharmaceutical point of view. Selection of salt forms of weak acids and weak bases based on their in vitro dissolution and in vivo PK properties will be discussed in the subsequent chapter with the emphasis on establishing potential in vitro and in vivo correlation. Earlier work on pharmaceutical salts, their dissolution rate, and its impact on bioavailability have been nicely reviewed by Berge et al. (63).

The relationship between diffusion layer pH and dissolution was first demonstrated by Nelson (64), whereas Mooney et al. (58) discussed the relationship with pH of the unbuffered bulk medium. A direct correlation has also been established between dissolution rates versus diffusion layer pH for various acidic and basic drugs (65). A self-buffering effect to maintain the steady-state microenvironmental pH may vary from compound to compound, depending on its solubility and/or pK_a values. Figure 12 shows the effect of the pH of dissolution medium on the dissolution of three acidic compounds having similar pK_a values, namely benzoic acid, 2-naphthoic acid, and indomethacin (57). In this figure, the increase in dissolution rate (flux) relative to that of the unionized species (N/N_0)

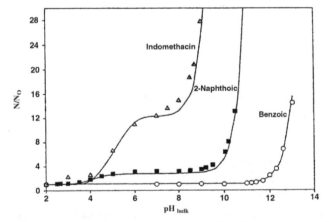

Figure 12 Relative flux (N/N_0) versus pH bulk for several carboxylic acids at 25°C. N_0 is the respective flux at pH 2. *Source*: From Ref. 57.

has been plotted. Despite similar pKa and pH-dependent solubility of all three compounds studied, pH of dissolution medium (bulk pH) had minimal effect on the dissolution of benzoic acid and 2-naphthoic acid, whereas a more pronounced effect was noticed in the dissolution of indomethacin. Indomethacin has much lower solubility values at different pH than that of other two acids and, as a result, its self-buffering effect in the diffusion layers is limited. The buffer capacity of a dissolution medium also plays a critical role in modulating micro-environmental pH of a drug substance and, therefore, its dissolution rate, as demonstrated earlier in Figure 8.

Salts of Weak Acids

The literature reports on human PK profiles of several weak acids and their salts have been reviewed. An example of the salts of weak acids is described herewith.

When the PK profiles of ibuprofen and its lysinate salt are compared (Fig. 13) (66), the lysinate salt demonstrates a shorter T_{max} and higher C_{max}, whereas the overall AUC was not altered significantly. Similarly, absorption of naproxen and its sodium salt resembles that of ibuprofen (Fig. 14). Interestingly, when commercial tablets of naproxen (Naprosyn) and naproxen sodium (Anaprox) were compared for its in vitro release at different pHs in the authors' lab, comparable dissolution profiles were obtained at pH 2.0 and 6.5. Dissolution at acidic pH (pH 2.0) of both forms is low due to their limited intrinsic solubility, whereas dissolution at neutral pH (pH 6.5) is controlled by

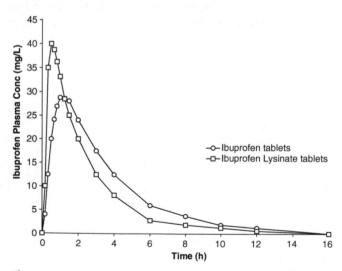

Figure 13 Mean plasma ibuprofen concentration-time profiles in 26 healthy volunteers following administration of 2×200 mg of ibuprofen as ibuprofen lysinate tablets (Nurofen Advance®) or sugar-coated tablets (Advil®). *Source*: From Ref. 66.

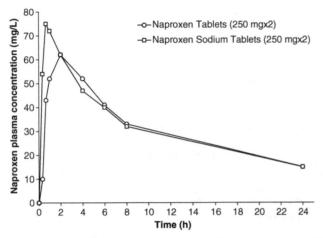

Figure 14   Comparative absorption of naproxen (Naprosyn^{TM}) and naproxen sodium (Anaprox^{TM}).

both the dissolution and disintegration of the dosage form. However, when a dissolution test using a pH switch method to simulate in vivo pH gradient is used, the advantage of rapid dissolution rate of the naproxen sodium versus that of naproxen is evident (Fig. 15). The dissolution test is performed at pH 2 for 30 minutes followed by pH ramp to 6.5 for 30 minutes by addition of concentrated sodium phosphate, then the pH is ramped up to 7.4. A rank order IVIVC can be established in this case for a weak acid (naproxen) and its salt (naproxen sodium). When salt of the weak acid is dissolved at acidic pH, conversion to free acid may occur, however, the acid coated particle or fine precipitates of acid/salt mixture dissolve rather rapidly when the pH is switched to intestinal pH. The pH gradient dissolution method may be a useful tool at early stage of development when comparing performance of different salt forms.

Salts of Weak Bases

Prediction of the in vivo performance of weak bases and their salts could be more challenging due to the kinetic nature of the dissolution/precipitation process. In another example, a weak basic drug NVS-1 containing different salt forms of the weak base were intravenous (IV) tested in a dog bioavailability study. The formulations tested are intraveneous formulation containing 3 mg/mL diHCl salt in 20% HP-β-CD and oral formulations, including free base in 0.5% carboxymethylcellulose (CMC) suspension at 2 mg/mL as well as diHCl and tartrate salts, which are in a dry blend capsule form using generic IR excipients (Table 2).

　　The in vivo PK profiles of the three NVS-1 oral variants were tested in three dogs and the results are summarized in Figure 16. Absolute bioavailability of the diHCl salt has a mean value of 84%, indicating close to complete absorption.

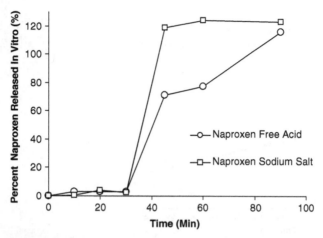

Figure 15 Dissolution profile of naproxen free acid and naproxen sodium salt commercial tablets using a pH gradient method (pH 2 → pH 6.5 at 30 minutes, pH 6.5 → pH 7.4 at 60 minutes).

Percentage bioavailability of the salt forms and free base suspension are in the order of diHCl (84%) > tartrate (48%) > free base (12%). Two salt forms of NVS-1 clearly demonstrate their in vivo advantages over that of the free base.

When in vitro dissolution profiles of the three per oral (p.o.) variants (i.e., free base suspension, diHCl salt, and tartrate salt IR capsules) were performed at pH 2 (Fig. 17), comparable release profiles are obtained for diHCl and tartrate salt of NVS-1, whereas release of the free base suspension was low (<20%) due the

Table 2 Summary of Formulations Administered to Dogs

Route	Dose[a] (mg/kg)	NVS-1	Volume/ No. of capsules	Concentration	Formulation
IV	3	diHCl salt	1 mL/kg	3 mg/mL	Solution in 20% hydroxypropyl-beta-cyclodextrin aqueous solution
p.o.	10	Free base	5 mL/kg	2 mg/mL	Suspension in 0.5% CMC aqueous solution
p.o.	10	diHCl salt	1 cap/dog	Not applicable	Powder-in-capsule
p.o.	10	Tartrate salt	1 cap/dog	Not applicable	Powder-in-capsule

[a]All doses are expressed as free base equivalent.
Abbreviations: IV, intravenous; CMC, carboxymethyl cellulose; p.o., per oral.

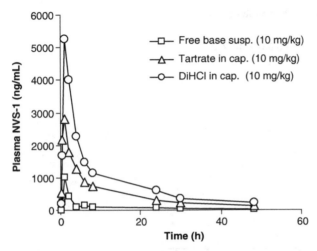

Figure 16 In vivo pharmacokinetic profiles of NVS-1 formulations (free base, tartrate salt, and diHCl salt).

limited solubility of free base and its ability to modulate microenvironment pH of the diffusion layer to a higher value.

Given the in vivo difference observed between the diHCl, tartrate, and free base suspension formulation, it warrants an in vitro method that could reflect the potential in vivo performance (i.e., a biorelevant method). When the dissolution of these formulations is tested at different pH levels, distinct different dissolution profiles are evident between the free base suspension, diHCl, and tartrate salt at pH 4 (Fig. 17). When the percent dissolved at pH 2 and pH 4 were compared with C_{max} and AUC of these formulations, it was evident that pH 4 provides a close to linear correlation between the in vitro percent release at 60 minutes and in vivo AUC or C_{max} (Fig. 18). Therefore, pH 4 should be used as an in vivo performance

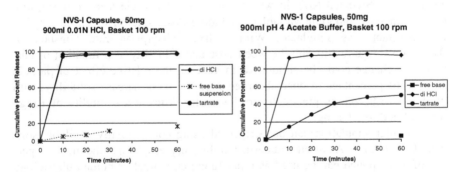

Figure 17 Dissolution profiles of NVS-1–diHCl, tartrate, and free base suspension at pH 2 and pH 4 media.

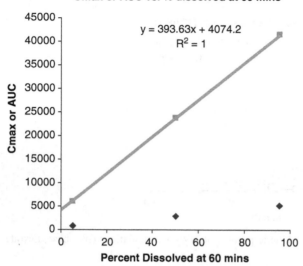

Figure 18 Level C correlation of C_{max} and area under the concentration time curve versus percent dissolution at 60 minutes. *Abbreviation*: AUC, area under the concentration time curve.

indicating method. The above data indicating Level C correlation can be established for NVS-1 using dissolution at pH 4. No further optimization was performed, since NVS-1 is only at preclinical stage. The dissolution method can be further optimized based on deconvoluted profiles, from human PK study.

Effect of Particle Size

The effect of particle size on in vivo drug absorption has been discussed by Johnson and Swindell (67), in which the effect of particle size on absorption over a range of important variables, including dose, solubility, and absorption rate constant was simulated. With a fixed absorption rate constant of 0.001 min^{-1}, the relationship between dose and solubility as a function of particle size change could be simulated. In general, the relative effect of particle size on the percentage of dose absorbed decrease with increase solubility, with particle size becoming practically irrelevant for drugs at a solubility of 1 mg/ mL with a dose of 1 mg.

Similar simulation can be performed using commercially available software GastroPlus®. As discussed earlier in this chapter, physicochemical properties of a compound can be used as input function, animal or human physiology can be selected. Simulation is not meant to be a replacement for scientific experiments, rather, it provides valuable insight on what one would expect to obtain in

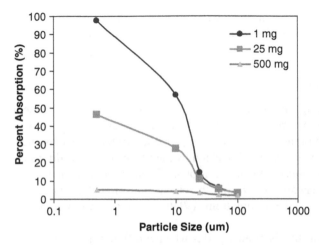

Figure 19 Simulation of absorption of a hypothetical compound with solubility at pH 7 = 1 μg/mL at doses of 1, 25, and 100 mg over particle size range of 0.5 to 100 μm.

vivo based on the physicochemical properties of a compound. For instance, when a compound with intrinsic solubility of 1 μg/mL at neutral pH and high effective permeability (3.0×10^{-4} cm/sec) is being tested, a simulation at different dose and different particle size could provide valuable information on the best possible scenario one could expect from such a compound (Fig. 19). At a dose of 500 mg, absorption of the compound is practically negligible over the particle size range of 500 nm to 100 μm, indicating that solubility limits the absorption of the compound; particle size reduction would not be helpful in such case to improve absorption. However, at dose of 1 mg, a dramatic shift in absorption dependency from solubility to particle size is evident from the simulation. Here, the use of micronization or nanoparticle system may provide great advantage.

Although one may not be able to obtain an accurate estimate of dose of a new chemical entity until very late in development, formulation scientists could utilize this type of information in a number of ways. At a relatively low dose range, relying on the particle size reduction and improving the wetting properties of a molecule may be quite effective. However, if the molecule is ionizable, such as a weak base, one may want to choose an alternative salt form, which could provide much higher solubility and dissolution rates compared with that of a free base. This would mitigate some of the risk of a potential dose increase at a later stage and, in addition, in vivo variability associated with variability coming from dissolution of the free form can also be minimized.

One key task during the drug development phase is to set a particle size specification. Such a specification is often set based on the impact of change in particle size on processibility or bioavailability. Using the example given earlier in the chapter (Fig. 11), a change in particle size from submicron to micron range does not appear to have any significant impact on the percentage of absorption of the molecule. This provides an invaluable tool to assist one to focus the research efforts on the impact of particle size on processibility rather than bioavailability.

Effect of Different Formulations

Physical properties of drug substance will often change during the development process as the chemists optimize their process chemistry, this could result in change in particle size, morphology, and surface properties of the drug substance. A drug product strategy, when greatly impacted by the properties of the drug substance, often needs to be adjusted to ensure a robust formulation that can be produced. A question that may arise is whether the change in formulation will result in bioinequivalence between the formulations. IVIVC can be a very powerful tool in such a case.

One of the questions that is often asked is whether different in vitro release properties will result in different in vivo absorption. It is not uncommon that different formulation approaches such as dry blend/direct compression, dry granulation (roller compaction), or wet granulation may result in formulations with different in vitro release characteristics. As shown in Figure 20, NVS-2 has different in vitro release profiles when dry blend and wet granulation formulations are tested in vitro. In early time points, the difference can be as large as 30%. However, when these formulations are tested in vivo in dog, comparable PK profiles are achieved (Fig. 21), indicating that the difference in the in vitro dissolution will not have any significant impact on the in vivo performance of the drug product. This would save an enormous amount of time and effort by avoiding the development of a process to match the dissolution profile of the original formulation.

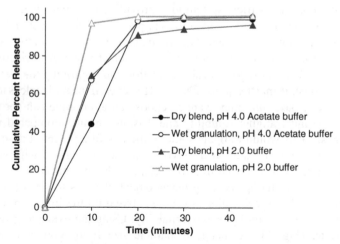

Figure 20   In vitro dissolution profiles of NVS-2 at pH 2.0 and pH 4.5.

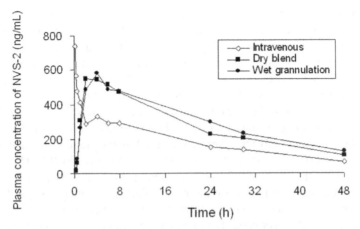

Figure 21 In vivo pharmacokinetic profiles of NVS-2: wet granulation and dry blend formulations in dogs.

IN VITRO–IN VIVO CORRELATIONS IN FULL DEVELOPMENT AND LIFE CYCLE MANAGEMENT

IVIVC can be even more useful at the later development stages or during life cycle management, where a number of human PK study have been performed at different doses and, sometimes, on different formulations. Dedicated discussion will be given to IR products in the following section, since IVIVC on MR formulations will be discussed by others.

The first group of IR products can be described by dissolution of not less than 85% within 30 minutes, and the dose of less than 250 mL of dissolution medium times lowest solubility in vivo relevant pH value range of one to eight. Since it can be assumed that these forms are already completely dissolved under in vivo conditions, all doses are available as a solution for absorption. Such rapid release forms are no longer different from a solution form, in terms of its PK behavior. In addition, if the compound has high permeability, then one could seek a biowaiver according to regulatory guidelines, since it is a BCS class I compound (68–70).

The second group consists of conventional IR products that display dissolution-limited absorption. IVIVC are possible with this group, so long as the absorption is dissolution-rate limited rather than permeation-rate limited. Substances from this group, which are mostly classified as class II by the BCS, are suitable candidates for IVIVC (71,72), and will be discussed in the following case study.

Immediate Release Product with Solubility-Limited Absorption

This case study refers to a proprietary compound characterized by solubility-limited absorption. To demonstrate the procedure of developing an IVIVC and for a better understanding, a few adjustments were made to the data, which serve to simplify and clarify the presentation of the IVIVC. The method presented in the previous sections for developing an IVIVC will be utilized; the IVIVC will be obtained based on conventional deconvolution using Wagner–Nelson method.

For NVS-3, dissolution tests during the formulation development were performed at 37°C with the USP apparatus 2 (paddle) in 900 mL of water and with the addition of 0.7% sodium lauryl sulfate using a rotational speed of 75 rpm. Since the NVS-3 drug substance showed solubility-limited absorption, USP apparatus 4 (flow through cell) was alternatively used as dissolution test method. The medium in the first hour was 0.1 N hydrochloric acid followed by 0.05 M phosphate buffer pH 6.8 with a flow rate of 16 mL/min and a flow cell volume of 20 mL.

Formulation screening resulted in formulation B, which was produced by roller compaction. Alternatively, a formulation A, which was produced by wet granulation, was also produced. The latter formulation shows a slower dissolution rate and the question is whether similar or different in vivo result would be obtained for these formulations.

The formulations (A vs. B) were tested in a single dose, crossover bioavailability study with 12 volunteers under fasting conditions. Blood samples were taken up to 72 hours after the dose was administered.

The plasma concentration-time curves are distinctly different with lower C_{max} and AUC values for formulation A compared to B, as illustrated in Figure 22. The in vivo absorption profiles of the tablets are calculated by the Wagner–Nelson method, the resulting absorption profiles are presented in

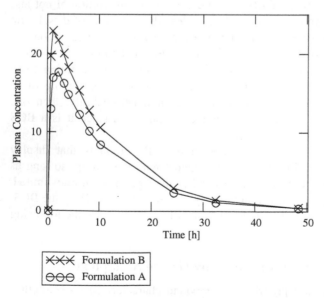

Figure 22 Plasma concentration profiles of immediate release formulations A and B.

Figure 23. The in vitro dissolution of both formulations was evaluated using both the paddle apparatus and the flow through cell apparatus (Fig. 24).

In order to determine the shape factor β, the in vivo and in vitro curves of formulation B were fit to a Weibull function utilizing least squares minimization. As mentioned earlier, only the curves having the same shape and, therefore, having the same β can be brought to a congruence, which will result in a successful IVIVC. As a consequence, the in vitro dissolution profile of formulation B should show a similar β compared with the β value of 0.38 of the in vivo dissolution profile.

The paddle method provides a shape factor of 0.53 (Fig. 24) that is distinctly closer to the in vivo factor than the shape factor of 1.28 obtained with the flow through cell method, and was therefore considered to be a more appropriate method. The operating conditions of the paddle method were then optimized using formulation B in order to find an in vitro profile for formulation B whose β is most similar to the β of the in vivo profile of this formulation. A 3-level full factorial design was chosen varying the rotation speed (50, 62, 75 rpm) and the concentration of sodium dodecyl sulfate (SDS) (0.4, 0.7, 1.0%). Figure 25 shows the results of the study in the contour plot of the corresponding modeling analysis. The plot illustrates the dependence of β from SDS-concentration and rotation speed. Each connected black line represents operating conditions that provide the same shape factor. The triangles symbolize the nine operating conditions of the study, the italic values represent the β-values

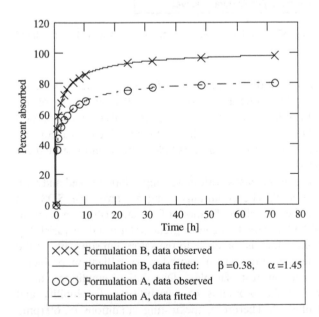

Figure 23 Absorption profiles of immediate release formulations A and B.

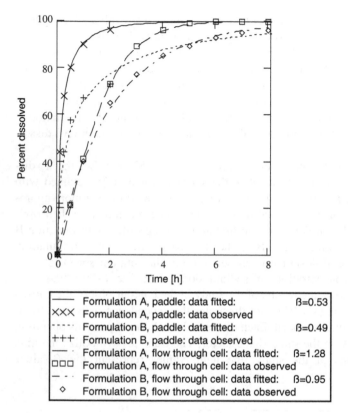

Figure 24 Dissolution profiles of immediate release formulations A and B with related β-values.

determined in the study. The bold dotted line shows the target value of 0.38, the squares represent the best three operating conditions. The goodness of fit, expressed as r^2, is 0.98. In order to verify the modeling analysis a dissolution test using 60 rpm/0.6% SDS was performed. The predicted value of β according to the model was 0.42 and the observed value was 0.44, which demonstrated the predictive power of the model.

The analysis shows that β increases with increasing rotation speed and with increasing concentration of SDS. The conditions using 50 rpm/1.0% SDS and 60 rpm/0.4% SDS provide the best shape factors of 0.36 and 0.40, compared to the in vivo target factor of 0.38; but the first causes great tablet-to-tablet variability and the latter uses a medium, which is too close to the saturation solubility of the drug substance at room temperature, which caused recrystallization problems during assay of the dissolution test samples. Dissolution conditions using 60 rpm/0.6% SDS overcame the problems of variability and solubility and provide a satisfactory β of 0.42. Therefore, the testing conditions of 60 rpm/0.6% SDS were chosen as operating conditions of the final dissolution test method.

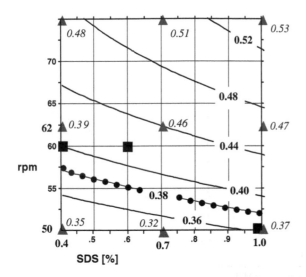

Figure 25 Contour plot of the modeling analysis for the dependence of β from sodium dodecyl sulfate concentration and rotation speed. *Abbreviation*: SDS, sodium dodecyl sulfate.

In order to bring the in vivo and in vitro profile in an optimum congruence, the in vitro profiles were scaled by a time scaling factor. The time scaling factor of 0.14 was determined by means of Equation 4 with an $\alpha_{\text{in vivo}}$ and $\alpha_{\text{in vitro}}$ of 1.44 and 0.21, respectively. This time factor was applied to the dissolution profiles of both formulations. Figure 26 presents the time-scaled in vitro profiles in comparison with the in vivo absorption profiles. In the initial phase up to 20 hours the curves are in good congruence, but thereafter the simulated profiles lie significantly above the observed profiles. In order to overcome this overestimation of absorption a cut-off factor was introduced, which keeps the profiles at the same plateau after 20 hours. By doing so there is a good congruence between the simulated and observed in vivo profiles.

The time-scaled and cut-off factor truncated dissolution profiles as well as the in vivo absorption profiles were used for linear regression. The result is illustrated in Figure 27 together with the formula of the regression line.

The relationship between the in vitro dissolution profile and the in vivo absorption profile in accordance with Equation 1, where the lag time t_0 with a value of 0 is not given in the parameters of the Equation, is as follows:

$$X_{\text{vivo}}(t) = 7.02 + 0.88 \times X_{\text{vitro}}(0.14 \times t) \quad \text{if } t > 20 \text{ then } t = 20 \tag{8}$$

In order to validate the IVIVC model, the internal predictability was calculated according to regulatory guidelines. The internal prediction errors for AUC are -4.0% for formulation A and 1.5% for formulation B and for C_{max} 3.2% and 3.1%, respectively.

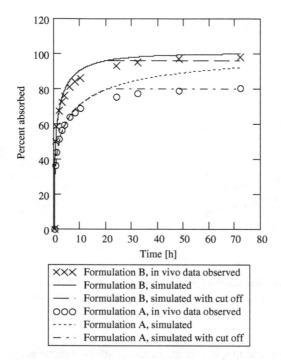

Figure 26 Absorption profiles of immediate release formulations A and B.

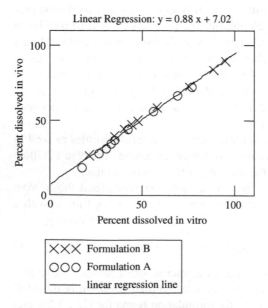

Figure 27 Linear regression of percent absorbed versus modified percent dissolved for immediate release formulations A and B.

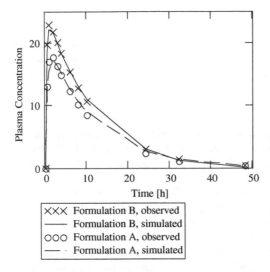

Figure 28 Simulated plasma concentration profiles of immediate release formulations A and B compared with observed data.

Figure 28 demonstrates the successful IVIVC by comparison of the simulated plasma concentration profiles with the observed profiles.

The initial issue raised in connection with the in vivo relevance of the difference in the dissolution profiles of both formulations can therefore be answered: a slower in vitro dissolution results in a slower in vivo dissolution with a resulting decrease in AUC.

CONCLUSIONS

IVIVC can be a very useful tool through the various product development stages. In a broad sense, IVIVC can be initiated by in silico simulation at the drug discovery stage. A more experimental based IVIVC with preclinical PK data and physicochemical data of the molecule is also possible. In the preclinical stage salt forms, particle sizes, and different formulations may also be evaluated. Further, human PK testing is an indispensable part of IVIVC development for the purpose of validating the IVIVC model. Last but not the least, IVIVC can be a very useful tool for projects at the life cycle management stage.

REFERENCES

1. Venkatesh S, Lipper RA. Role of the development scientist in compound lead selection and optimization. J Pharm Sci 2000; 89:145–154.
2. Lipper RA. How can we optimize selection of drug development candidates from many compounds at the discovery stage? Mod Drug Discov 1999; (Jan/Feb):55–60.

3. Li S, He H, Parthiban LJ, Yin H, Serajuddin ATM. IV-IVC considerations in the development of immediate-release oral dosage form. J Pharm Sci 2005; 94(7):1397–1417.
4. Center for Drug Evaluation and Research (CDER). Guidance for industry: Extended Release Oral Dosage Forms: Development, Evaluation, And Application of In Vitro/In Vivo Correlation. Rockville, MD: US Department of Health and Human Services, 1997.
5. Young D. In vitro in vivo correlation of controlled release parenterals, AAPS Meeting, Baltimore, MD, 2003.
6. Note for guidance on quality of modified release products: A: Oral Dosage Forms, B. Transdermal Dosage Forms, Section I (Quality). The European Agency for the Evaluation of Medicinal Products, Human Medicines Evaluation Unit (EMEA), London, 1999.
7. Food and Drug Administration, Center for Drug Evaluation (CDER). Guidance for industry: Modified Release Solid Oral Dosage Forms: Scale-up and Postapproval Changes: Chemistry, Manufacture and Controls, In Vitro Dissolution Testing and In Vivo Bioequivalence Documentation. September 1997.
8. Mahayni H, Rekhi GS, Uppoor RS. Evaluation of external predictability of an in vitro–in vivo correlation for an extended-release formulation containing metoprolol tartrate. J Pharm Sci 2000; 89(10):1354–1361.
9. United States Pharmacopeia XXVIII, General Information: In Vitro and In Vivo Evaluation of Dosage Forms. Ch. 1088, 2005.
10. Turner S, Federici C, Hite M, et al. Formulation development and human in vitro–in vivi correlation for a novel, monolithic controlled-release matrix system of high load and highly water-soluble drug niacin. Drug Dev Ind Pharm 2004; 30(8):797–807.
11. Brockmeier D. In vitro/in vivo correlation of dissolution using moments of dissolution and transit times. Acta Pharm Technolo 1986; 32(4):164–174.
12. Podczeck F. Comparison of in vitro dissolution profiles by calculating mean dissolution time (MDT) or mean residence time (MRT). Int J Pharm 1993; 97:93–100.
13. Lake OA, Olling M, Barends DM. In vitro/in vivo correlations of dissolution data of carbamazepine immediate release tablets with pharmacokinetic data obtained in healthy volunteers. Eur J Pharm Biopharm 1999; 48:13–19.
14. Abrahamsson B, Lennernäs H. Application of the biopharmaceutic classification system now and in the future. In: van de Waterbeemd H, Lennernäs H, Artursson P, eds. Drug Bioavailability, 1st ed. Weinheim: Wiley-VCH, 2003:495–531.
15. Drewe J, Guitard P. In vitro-in vivo correlation for modified formulations. J Pharm Sci 1993; 82(2):132–137.
16. Tashtouch BM, Al-Qashi ZS, Najib NM. In vitro and in vivo evaluation of glibenclamide in solid dispersion systems. Drug Dev Ind Pharm 2004; 30(6):601–607.
17. Posti J, Katila K, Kostiainen T. Dissolution rate limited bioavailability of flutamide, and in vitro-in vivo correlation. Eur J Pharm Biopharm 2000; 49:35–39.
18. Vaughan DP, Leach RH. Simple transformation method for predicting plasma drug profiles from dissolution rates. J Pharm Sci 1976; 65(4):601–603.
19. Shepard T, Hayes S, Farrell C. Development of a level A IVIVC for a BCS Class 2 IR product. http://www.globomaxservice.com/resource/AAPS2003_DevelLevelA.pdf (accessed August 2005).
20. Kortejaervi H, Mikkola J, Baeckman M, Antila S, Marvola M. Development of level A, B and C in vitro-in vivo correlations for modified-release levosimendan capsules. Int J Pharm 2002; 241(1):87–95.

21. Akikmoto M, Furuya A, Nakamura M. Release and absorption characteristics of chlorphenesin carbamate sustained-release formulations: in vitro-in vivo and in vivo dog-humen correlations. Int J Pharm 1995; 117:31–39.
22. Rao BS, Seshasayana A, Saradhi SVP. Correlation of in vitro release and in vivo absorption characteritics of rifampicin from ethylcellulose coated nonpareil beads. Int J Pharm 2001; 230:1–9.
23. Langenbucher F. Handling of computational in vitro/in vivo correlation problems by Microsoft Excel II. Principles and some general algorithms. Eur J Pharm Biopharm 2002; 53:1–7.
24. Wagner JG. Application of the Loo-Riegelman absorption method. J Pharmacokin Biopharm 1975; 3(1):51–67.
25. Frick A, Möller H, Wirbitzky E. Biopharmaceutical characterization of oral controlled/modified-release drug Products. In vitro/in vivo correlation of roxatidine. Eur J Pharm Biopharm 1998; 46:313–319.
26. Langenbucher F. Numerical convolution/deconvolution as a tool for correlating in vitro with in vivo drug availability. Pharm Ind 1982; 44:1166–1172.
27. Gibaldi M, Perrier D. Pharmacokinetics. 2nd ed. New York: Marcel Dekker, 1982.
28. Shargel L, Yu ABC. Applied Biopharmaceutics & Pharmacokinetics. 4th ed. East Norwalk, CT: Appleton-Century-Crofts, 1999.
29. Humbert H, Cabiac MD, Bosshardt H. In vitro-in vivo correlation of a modified-release oral form of ketotifen: in vitro dissolution rate specification. J Pharm Sci 1994; 83(2):131–136.
30. Langenbucher F. Handling of computational in vitro/in vivo correlation problems by Microsoft Excel: III. Convolution and deconvolution. Eur J PharmBiopharm 2003; 56(3):429–437.
31. Langenbucher F. Handling of computational in vitro/in vivo correlation problems by Microsoft Excel: IV. Generalized matrix analysis of linear compartment systems. Eur J Pharm Biopharm 2005; 59(1):229–235.
32. Buchwald P. Direct, differential-equation-based in vitro–in vivo correlation (IVIVC) method. J Pharm Pharmacol 2003; 55(4):495–504.
33. Gillespie WR. Convolution-based approaches for in vivo–in vitro correlation modeling. In: Young D, Devane JG, Butler J. eds. In Vitro–In Vivo Correlations. New York: Plenum. Adv Exp Med Biol 1997; 423:53–65.
34. Gomeni R, Angeli C, Bye A. In silico prediction of optimal in vivo delivery properties using convolution-based model and clinical trial simulation. Pharm Res 2002; 19(1):99–103.
35. Nicolaides E, Symillides M, Dressman JB. Biorelevant dissolution testing to predict the plasma profile of lipophilic drugs after oral administration. Pharm Res 2001; 18(3):380–388.
36. Polli JE, Rekhi GS, Augsburger LL, et al. Methods to compare dissolution profiles and a rationale for wide dissolution specifications for metoprolol tartrate tablets. J Pharm Sci 1997; 86(6):690–700.
37. Shepard T, Farrell C. The incorporation of time-scaling in ivivc models to account for apparent non-linear relationships, http://www.globomaxservice.com/resource/M1226.pdf (accessed August 2005).
38. Shepard T, Rochdi M. development and validation of a time scaled in vitro/in vivo correlation for isosorbide-5-mononitrate controlled release geomatrix formulations,

http://www.globomaxservice.com/resource/M1350_final.pdf (accessed August 2005).

39. Drewe J, Guitard P. In vitro-in vivo correlation for modified formulations. J Pharm Sci 1993; 82(2):132–137.
40. Liu Y, Schwartz JB, Schnaare RL, et al. A multi-mechanistic drug release approach in a bead dosage form and in vivo predictions. Pharm Dev Technol 2003; 8(4):409–417.
41. Huang Y-B, Tsai Y-H, Yang W-C, et al. Once-daily propranolol extended-release tablet dosage form: formulation design and in vitro/in vivo investigation. Eur J Pharm Biopharm 2004; 58:607–614.
42. Corrigan OI, Devlin Y, Butler J. Influence of dissolution medium buffer composition on ketoprofen release from ER products and in vitro-in vivo correlation. Int J Pharm 2003; 254(2):147–154.
43. Takka S, Sakr A, Goldberg A. Development and validation of an in vitro–in vivo correlation for buspirone hydrochloride extended release tablets. J Control Release 2003; 88:147–157.
44. Qiu Y, Garren J, Samara E, et al. Once-a-day controlled-release dosage form of Divalproex Sodium II: development of a predictive in vitro drug release method. J Pharm Sci 2003; 92(11):2317–2325.
45. Polli JE, Crison JR, Amidon GL. Novel approach to the analysis of in vitro–in vivo relationships. J Pharm Sci 1996; 85:753–760.
46. Dunne A, O'Hara T, DeVane J. Level A in vivo–in vitro correlation: nonlinear models and statistical methodology. J Pharm Sci 1997; 86(11):1245–1249.
47. Lipinski CA. Computational alerts for potential absorption problems: profiles of clinically tested drugs. Tools for oral absorption. Part Two. Predicting human absorption. Biotech, PDD symposium, AAPS, Miami, 1995.
48. Palm K, Luthman K, Ungell AL, Strandlund G, Artursson P. Correlation of drug absorption with molecular surface properties. J Pharm Sci 1996; 85(1):32–39.
49. Perez MA, Sanz MB, Torres LR, Avalos RG, Gonzalez MP, Diaz HG. A topological sub-structural approach for predicting human intestinal absorption of drugs. Eur J Med Chem 2004; 39(11):905–916.
50. Hall LH, Hall LM. QSAR modeling based on structure-information for properties of interest in human health. SAR QSAR Environ Res 2005; 16(1–2):13–41.
51. Bai JPF, Utis A, Crippen G, et al. Use of classification regress tree in predicting oral absorption in humans. J Chem Inf Comput Sci 2004; 44:2061–2069.
52. Obata K, Sugano K, Saitoh R, et al. Prediction of oral drug absorption in humans by theoretical passive absorption model. Int J Pharm 2005; 293(1–2):183–192.
53. Wolohan PR, Clark RD. Predicting drug pharmacokinetic properties using molecular interaction fields and SIMCA. J Comput Aided Mol Des 2003; 17(1):65–76.
54. Yu LX, Crison Jr, Amidon GL. Compartmental transit and dispersion model analysis of small intestinal transit flow in humans. Int J Pharm 1996; 140:111–118.
55. Parrot N, Lave T. Prediction of intestinal absorption: comparative assessment of GASTROPLUS and IDEA. Eur J Pharm Sci 2002; 17:51–61.
56. Horter D, Dressman JB. Influence of physicochemical properties on dissolution of drugs in the gastro-intestinal tract. Adv Drug Deliv Rev 2001; 46:75–87.
57. Ozturk SS, Palsson BO, Dressman JB. Dissolution of ionizable drugs in buffered and unbuffered solutions. Pharm Res 1988; 5:272–282.

58. Mooney KG, Mintun MA, Himmelstein KJ, Stella VJ. Dissolution kinetics of carboxylic acids I: effect of pH under unbuffered conditions. J Pharm Sci 1981; 70:13–22.
59. Serajuddin ATM, Jarowski C. Effect of diffusion layer pH and solubility on the dissolution rate of pharmaceutical acids and sodium salts II: salicylic acid, theophylline, and benzoic acid. J Pharm Sci 1985; 74:148–154.
60. Morris KR, Fakes MG, Thakur AB, et al. Integrated approach to the selection of optimal salt form for a new drug candidate. Int J Pharm 1994; 105:209–217.
61. Anderson BD, Flora KP. Preparation of water-soluble compounds through salt formation. In: Wermuth CG, ed. The Practice of Medicinal Chemistry. London: Academic Press, 1996:739–754.
62. Gould PL. Salt selection for basic drugs. Int J Pharm 1986; 33:201–217.
63. Berge SM, Bighley LD, Monkhouse DC. Pharmaceutical salts. J Pharm Sci 1977; 66:1–19.
64. Nelson E. Solution rate of theophylline salts and effects from oral administration. J Am Pharm Assoc Sci Educ 1957; 46:607.
65. Serajuddin ATM, Jarowski CI. Effect of diffusion layer pH and solubility on the dissolution rate of pharmaceutical bases and their hydrochloride salts. Part 1. Phenazopyridine. J Pharm Sci 1985; 74:142–147.
66. Schettler T, Paris S, Pellett M, Kidner S, Wilkinson D. Comparative pharmacokinetics of two fast-dissolving oral ibuprofen formulations and a regular release ibuprofen tablet in healthy volunteers. Clin Drug Invest 2001; 21(Jan): 73–78.
67. Johnson KC, Swindell AC. Guidance in the setting of drug particle size specifications to minimize variability in absorption. Pharm Res 1996; 13:1795–1798.
68. Food and Drug Administration, Center for Drug Evaluation (CDER), Rockville, MD. Guidance for industry, waiver of in vivo bioavailability and bioequivalence studies for immediate-release solid oral dosage forms based on a biopharmaceutics classification system, August 2000.
69. Fassen F, Vromans H. Biowaivers for oral immediate-release products. Clin Pharmacokinet 2004; 43(15):1117–1126.
70. Verbeek RK, Junginger HE, Midha KK. Biowaiver monographs for immediate release solid oral dosage forms based on biopharmaceutics classification system (BCS) literature data: chloroquine phosphate, chloroquine sulfate, and chloroquine hydrochloride. J Pharm Sci 2005; 94(7):1389–1395.
71. Veng-Pedersen P, Gobburu JV, Meyer MC, Straughn AB. Carbamazepine level-A in vivo in vitro correlation (IVIVC): scaled convolution based predictive approach. Biopharm Drug Dispos 2000; 21(1):1–6.
72. Sunesen VH, Pedersen BL, Kristensen HG, Muellertz A. In vivo in vitro correlations for a poorly soluble drug, danazol, using the flow-through dissolution method with biorelevant dissolution media. Eur J Pharm Sci 2005; 24(4):305–313.

5

In Vitro–In Vivo Correlation

Nishit B. Modi

Department of Clinical Pharmacology, ALZA Corporation, Mountain View, California, U.S.A.

INTRODUCTION

A key objective of pharmaceutical product development is a good understanding of the in vivo and in vitro performance of dosage forms. This understanding may be particularly valuable for controlled release dosage forms since it allows demonstration of consistent and predictable performance. A good in vitro–in vivo correlation (IVIVC) can allow the use of in vitro dissolution studies for production control and allows prediction of in vivo performance based on laboratory data. Numerous approaches have been used to develop IVIVCs (1,2).

DEFINITIONS

According to the U.S. Food and Drug Administration (FDA), an IVIVC is a predictive mathematical model describing the relationship between an in vitro property of an oral dosage form (usually the rate or extent of drug dissolution or release) and a relevant in vivo response (e.g., plasma drug concentrations or amount of drug absorbed) (3).

The U.S. Pharmacopoeia defines an IVIVC as "the establishment of a (quantitative) relationship between a biological property or a parameter derived from a biological property produced by a dosage form, and a physico-chemical property or characteristic of the same dosage form" (4). The biological properties most commonly used in developing IVIVCs are the maximum drug concentration (C_{max}) or area under the concentration-time curve (AUC), obtained from a clinical study. The physicochemical property most commonly used is the in vitro dissolution profile.

Unlike immediate-release dosage forms, controlled or modified release products are designed to provide a distinct concentration profile, generally extending beyond 12 hours, and cannot be characterized by a single-point dissolution test. For this reason, it is often easier to develop an IVIVC for a controlled-release product than it is for an immediate-release dosage form.

Successful development and application of an IVIVC requires that dissolution or release of drug from the dosage form be the rate-limiting step in the sequence of steps leading to drug absorption into the systemic circulation (5). As such, in some cases, IVIVCs may not be possible for immediate release formulations where dissolution is often not the rate-limiting step. A meaningful IVIVC would be of benefit as a surrogate for bioequivalence studies that might otherwise be required with scale-up or minor postapproval changes in formulation, equipment, manufacturing process, or in manufacturing site, and such an IVIVC might improve product quality and reduce regulatory burden (3,6).

LEVELS OF CORRELATION

Four levels of correlation (Levels A, B, C, and multiple Level C) have been described in the FDA guidance.

Level A

Level A correlations are the highest level of correlation, representing a point-to-point correlation between the in vitro input rate (e.g., dissolution rate) and the in vivo input rate. The correlation is usually linear, but nonlinear correlations are also possible (7). Level A correlations are considered the most informative and useful for developing novel dosage forms and for regulatory support. The typical process of developing a Level A IVIVC involves deconvolution of the in vivo plasma profile to estimate the in vivo release, followed by comparison of the in vivo fraction of drug absorbed to the in vitro fraction of drug dissolved. Deconvolution of the plasma profile may be done either using mass balance, model-dependent methods such as the Wagner-Nelson or Loo-Regelman methods, or by model independent, mathematical deconvolution (8–12). In the case of a linear relationship, the in vitro fraction dissolved and the in vivo fraction absorbed will be superimposable or may be made superimposable by the use of a scaling factor (or less commonly using a scaling function). For an IVIVC to be valid, a single scaling factor or scaling function should be applicable to different release rates of the same formulation. It is possible that alternative approaches can be used to develop valid IVIVCs. Regardless of the method, a validate Level A IVIVC should be robust, able to predict the entire dissolution time course (or plasma profile), and versatile enough to discriminate between different in vitro dissolution rate profiles.

A Level A correlation relates the entire in vitro dissolution profile to the in vivo concentration profile, and the in vitro dissolution can then serve as a surrogate of the in vivo performance. Such an IVIVC can subsequently be used as a quality control procedure, and potentially can support minor changes in manufacturing process, changes in raw material supplies, and manufacturing site changes without the need for additional clinical studies. Although many publications make a point of noting that Level A correlations are point-to-point correlations, it is more important to recognize that a level A correlation allows prediction of the entire in vivo concentration time course from the in vitro dissolution data.

Level B

Level B correlations use the principles of statistical moments to compare a summary parameter of the mean in vitro dissolution rate (e.g., mean dissolution time) and a mean in vivo summary parameter [e.g., mean residence time (MRT) or mean in vivo dissolution time]. Although Level B correlations use all the in vitro and in vivo data, they are derived using a single integrated parameter, and such correlations are not point-to-point correlations. Level B correlations are not considered very useful, since a variety of different in vitro and in vivo profiles can result in the same in vitro and in vivo summary parameters. Level B correlations are also not very useful for regulatory purposes since they do not reflect the full in vivo concentration-time profile.

Level C

A Level C correlation establishes a single point correlation between an in vitro dissolution parameter [e.g., the time to release 50% of the drug (T_{50})] and an in vivo parameter (e.g., C_{max} or AUC). Level C correlations also do not reflect the complete in vivo concentration time course, and hence are not very useful for regulatory support. Such correlations may be of some use during early formulation development.

Multiple Level C correlations extend the single point Level C correlation to relate several in vivo parameters to in vitro parameters related to drug release, at several time points of the dissolution profile. As such, Level C correlations can be as useful as Level A correlations. However, if a multiple Level C correlation is possible, it is likely that a Level A correlation can also be developed.

BIOPHARMACEUTICAL CLASSIFICATION SYSTEM AND IN VITRO–IN VIVO CORRELATION

The biopharmaceutical classification system (BCS) is a framework for classifying drugs based on their aqueous solubility and intestinal permeability (13). Table 1 summarizes the expectations for development of an IVIVC for immediate-release products based on the BCS.

Table 1 Expectation of IVIVC for Immediate Release Products Based on the Biopharmaceutical Classification System

BCS class	Characteristic	IVIVC expectation
I	High solubility/high permeability	IVIVC not likely
II	Low solubility/high permeability	IVIVC should be possible
III	High solubility/low permeability	IVIVC not likely
IV	Low solubility/low permeability	Low possibility of IVIVC

As noted previously, an IVIVC is likely when dissolution is the rate-limiting step and the drug has high permeability. BCS class I compounds have high solubility and high permeability. Immediate release formulations of class I compounds will be released in the stomach, and provided gastric absorption is limited, gastric emptying will be the rate limiting step—a process that is not accounted for by in vitro dissolution testing. When the in vivo dissolution rate is rapid in relationship to gastric emptying and the drug has a high intestinal absorption, in vitro dissolution is not likely to adequately reflect absorption and an IVIVC is not likely for class I compounds.

BCS class II compounds have a low solubility and a high permeability—in this case, absorption is likely to be dissolution rate-limited. Hence, for this class of drugs, an IVIVC is likely to be established.

BCS class III compounds have a high solubility and a low permeability. Like class I compounds, drug release is not dissolution rate-limited and an IVIVC is unlikely unless dissolution is slower than intestinal permeability. BCS class IV compounds have a low solubility and a low permeability, and consequently an IVIVC is unlikely.

CONSIDERATIONS IN DEVELOPING AN IN VITRO–IN VIVO CORRELATION

The typical process of developing a robust Level A IVIVC involves the following steps:

1. Development of formulations that are designed to release drug at different release rates. Three formulations designed to release drug at a slow, medium, and rapid rate may generally be sufficient. Typically, these lots should meet (or be close to) the anticipated limits of the in vitro release specifications.
2. In vitro dissolution data are generated using an appropriate dissolution test. FDA guidance indicates using at least 12 individual dosage forms from each lot, with sampling points that allow adequate characterization of the dissolution profile with a coefficient of variation of less than 10% for the mean dissolution profile.

3. The lots are included as part of a pharmacokinetic study to obtain in vivo plasma concentration data. Ideally, the lots are compared in a single crossover pharmacokinetic study with a sufficient number of subjects (typically 12–36 subjects). However, the comparisons may be done in parallel design studies or derived from several different studies that are conducted as part of the product development process. In any event, it is expected that the in vitro release profiles are sufficiently different, and correspondingly, the in vivo profiles are also different. In some instances, it may be very useful to include a reference to allow correction for inter-occasion variability. This may be an immediate release formulation (solution, suspension, or even a tablet) or an intravenous administration.

4. Finally, an appropriate method is used to estimate the in vivo absorption or dissolution time course. This can be done using compartmental methods such as the Wagner-Nelson or Loo-Riegelman method, or by noncompartmental methods such as numerical deconvolution. The inclusion of the reference treatment may be useful for deconvolution. Additionally, should the data available for developing an IVIVC come from different studies, a reference treatment may sometimes be useful for the purpose of normalizing differences in bioavailability between studies or accounting for inter-occasion variability.

5. During the development of an IVIVC it may be necessary to explore different in vitro dissolution conditions, to identify conditions that discriminate between formulations.

6. Generally, the formulations available are divided into two groups: one set of data that are used for development of an IVIVC and for estimating internal prediction error, and a group that is not used in the development of the IVIVC but is available for demonstrating external prediction. The goodness of fit for the IVIVC is assessed by the percentage prediction error estimated as follows:

$$\% \, \text{Prediction error (PE)} = \frac{\text{Observed value} - \text{Predicted value}}{\text{Observed value}} \times 100$$

7. If three or more release rates are incorporated in the development of an IVIVC, the mean absolute internal prediction error is less than 10% for C_{max} and for AUC_{inf}, and the %PE for individual release rates used in the IVIVC does not exceed 15%, the IVIVC is considered to have good predictability and no further assessment or validation is required. It may still be of value to assess external prediction of the IVIVC.

8. If the internal prediction error is greater than 10% and/or the %PE for an individual release rate exceeds 15%, FDA guidance requires external validation. For this validation, datasets that were not used in the development of the IVIVC are used to estimate the external %PE. If the mean absolute %PE is less than 10% for C_{max} and for AUC_{inf},

external predictability is demonstrated. Validation is considered inconclusive if the %PE is between 10% to 20%, and a %PE greater than
20% indicates inadequate validation.

The process of developing an IVIVC outlined earlier is illustrated in
Figure 1 using data for OROS® oxprenolol (14). In this published study, eight
healthy male volunteers received doses of OROS® oxprenolol, an intravenous
bolus injection, and immediate release tablets in a crossover fashion. Plasma
oxprenolol concentration data for the two lots of OROS® that were most dissim-
ilar (formulations A and E) were selected for demonstrating the process of developing an IVIVC. The cumulative in vitro release rates and corresponding mean in
vivo plasma oxprenolol profiles are shown in panels A and B, respectively. A bi-
exponential function was fit to the mean plasma concentration data following the
intravenous treatment. To estimate the in vivo release profile, the mean plasma
concentration time profiles following oral treatment of formulations A and E
were deconvolved using the intravenous treatment as the characteristic response.
A plot of the cumulative in vitro release against the in vivo release results in a
linear relationship, with a slope approximately equal to unity following

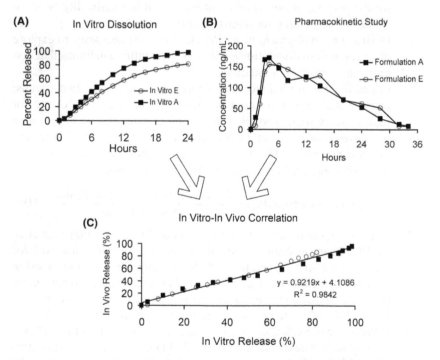

Figure 1 Schema for developing in vitro–in vivo correlation (IVIVC). (**A**) Cumulative
in vitro dissolution profile, (**B**) in vivo concentration time profile from a pharmacokinetic
study, and (**C**) IVIVC.

correction for differences in total drug release between the two formulations, demonstrating Level A IVIVC (panel C).

The earlier discussion focusses on a deconvolution approach, that is, it deconvolves the in vivo plasma profile to extract an in vivo release profile that may then be compared with the in vitro release (or dissolution) profile. It is also possible to apply a convolution-based approach to convolve the in vitro dissolution profile with a characteristic response (also referred to as a unit impulse), to predict an in vivo plasma profile that may be compared with the observed plasma profile. In a convolution-based approach the parameters estimated would be those related to the characteristic response (the pharmacokinetic parameters).

As noted previously, there are often instances where the base approach outlined earlier may not be adequate, and a correction may need to be applied. Figure 2 presents the plasma concentration time profiles for three formulations of an antiepileptic drug that were investigated as part of a formulation development program (panel A). The corresponding cumulative in vitro profiles are

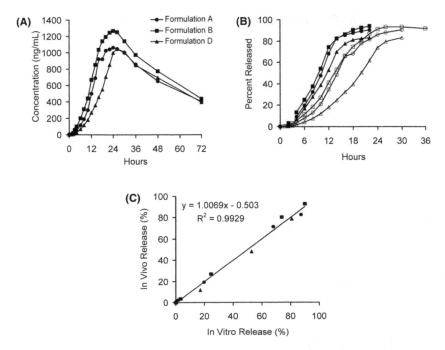

Figure 2 Use of scaling in development of a Level A in vitro–in vivo correlation. (**A**) In vivo concentration-time profile for three test formulations. (**B**) Comparison of in vitro (closed symbols) and in vivo (open symbols) cumulative release showing a systematic shift between the two. (**C**) Comparison of in vitro and in vivo release following application of a scalar correction to the in vivo profiles. Following the correction there is a good Level A correlation.

shown in panel B (solid symbols). To develop an IVIVC for the dosage form, the in vivo data were deconvolved using the Wagner-Nelson approach. The resulting cumulative in vivo profiles for the three formulations are also presented in panel B (open symbols). It is clear that while the cumulative amount released in vitro and in vivo are comparable, there is a systematic rightward time shift for the in vivo profile—there is an apparent in vivo slower release compared to the in vitro release. To correct for this, a scalar (constant) time shift was applied to the in vitro release data from all three formulations. Plotting the corrected in vitro and in vivo cumulative release data now results in a type A IVIVC with a linear relationship and a slope of unity (panel C).

To illustrate the principles of determining Level B and Level C correlations, the MRT values and mean dissolution time for the three formulations in the early mentioned antiepileptic example were estimated from the mean plasma concentration-time profiles and in vitro release profiles. Similarly, the AUC to the time to maximum concentration ($AUCT_{max}$) and the time for 50% of in vitro drug release were estimated. Figure 3 presents a plot of the mean

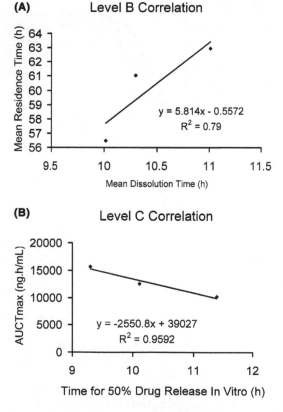

Figure 3   Development of Level B (**A**) and Level C (**B**) in vitro in vivo correlations.

dissolution time against the MRT (panel A) to illustrate a Level B correlation. Similarly, panel B (Fig. 3) shows a multiple Level C correlation obtained by plotting the time for 50% drug release in vitro against the $AUCT_{max}$ values. Although both correlations were reasonable, their utility is limited and they do not provide as much information as the Level A correlation.

REGULATORY CONSIDERATIONS

Setting Dissolution Specifications

The FDA guidance indicates that in the absence of an IVIVC, the recommended range of deviation for setting the upper (or lower) limit of the release specification at any point is 10% from the mean dissolution profile. Reasonable deviation from this may be acceptable provided that the range at any time point does not exceed 25%. A specification greater than 25% may be acceptable based on demonstration that the mean dissolution profiles allowed by the upper and lower limits of the specification are bioequivalent. Ideally, specifications should be established such that all lots within the upper and lower limit allowed are bioequivalent. The guidance states that as a less optimal option, lots at the proposed upper and lower limits of the dissolution specification should be bioequivalent to the clinical/bioavailability lots or to an appropriate reference standard. It is not evident how acceptable this latter option is, since it leaves open the possibility that lots at the upper and lower limits of the dissolution specifications may not be bioequivalent to each other.

When an IVIVC is available, its information can be used to define dissolution specifications. Generally, average data (in vitro dissolution and pharmacokinetic profile) are used, although more elaborate approaches can be used, which include variability in the dissolution and/or pharmacokinetic data or involving Monte Carlo simulations spanning the desired specification range. A minimum of three time points, covering the early, medium term, and late stages of the dissolution, is necessary to fully characterize the dissolution profile.

In setting dissolution specifications, the in vivo plasma profiles resulting from release profiles spanning the upper (fastest profile) and lower limits (slowest profile) of specifications are predicted through convolution or alternative means based on the IVIVC. The C_{max} and AUC_{inf} values for the two profiles are compared with each other and should not differ by more than 20% for acceptance of the proposed dissolution profile. Availability of a robust IVIVC may allow dissolution specifications with a range wider than the 25% allowed in the absence of an IVIVC, provided the slowest and fastest profiles result in AUC_{inf} and C_{max} values that are bioequivalent.

In some cases (e.g., for osmotic release formulations), it may be more appropriate to specify a release rate as a control parameter, either alone or in addition to the cumulative release specifications. This may be particularly important for formulations designed to provide a zero-order release for a specified duration or for formulations that are designed to provide a specific release rate. In such instances, an IVIVC allows definition of the interval where such a specification may be most relevant for controlling the process.

Consider the following example on the use of simulation and modeling in setting release-rate specifications. OROS® pseudoephedrine is a once-a-day modified-release tablet for oral administration of 240 mg pseudoephedrine hydrochloride for the temporary relief of nasal congestion due to the common cold, seasonal allergic rhinitis or other respiratory allergies, and nasal congestion associated with sinusitis. It has been demonstrated that there is a 1:1 (Type A) correlation between the in vitro and in vivo release rates for OROS® pseudo-ephedrine (15). To illustrate the application of a Level A correlation in setting release-rate specifications, a one-compartment pharmacokinetic model was fit to the plasma concentration versus time data from a clinical study, involving oral pseudoephedrine hydrochloride solution to estimate the characteristic response. To validate the IVIVC, pharmacokinetic parameters estimated for the oral solution treatment were convolved with the in vitro release-rate data from OROS® pseudoephedrine tablets that were included in a clinical studyl, to predict the in vivo pseudoephedrine plasma concentration-time profile. A comparison of the predicted in vivo profile and the observed profile demonstrated good correlation supporting validation of a Level A IVIVC. Two hypothetical in vitro release rates, designed to support the upper and lower limits of the desired release-rate specification, were generated. One lot was designed to support a cumulative release rate of approximately 65% at 14 hours and the other approximately 90% at 14 hours, providing an overall range of 25% between the upper and lower limits. The cumulative amounts for the proposed in vitro specifications are shown graphically in Figure 4 along with the in vitro release rate of the lot included in a clinical study.

To compare the validity of the proposed upper and lower limits of the in vitro specifications, in vivo concentration profiles were predicted for lots that skirted the proposed upper and lower limits using the Level A IVIVC and compared with the lot included in the in vivo study (representing the midpoint of the proposed range). Table 2 presents a comparison of the pharmacokinetic parameters for the lot used in the clinical study and lots at the upper and lower limits of the proposed in vitro specification.

With the exception of T_{max}, the ratios of C_{max} and of AUC_{inf} values for the test to reference (the clinical lot) for both the 65% and 90% systems were within 20% of each other. For modified release dosage forms, determination of T_{max} is often difficult and statistical treatment of T_{max} is less important than for C_{max} and for AUC_{inf}. Based on the published correlation between the in vitro and in vivo release rates for pseudoephedrine hydrochloride (HCl) and the results of simulations, a release-rate specification that supports a 25% range in the cumulative amount released may be proposed for this product.

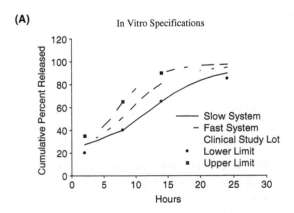

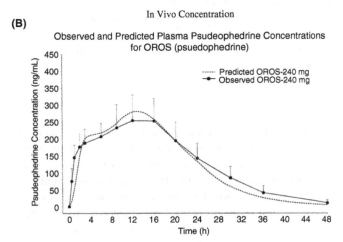

Figure 4 Application of in vitro–in vivo correlation for setting release specifications. (A) In vitro release profile for test lots and proposed in vitro release specifications. (B) Observed and predicted in vivo concentrations.

Table 2 Comparison of the Pharmacokinetic Parameters for OROS® Pseudoephedrine Lots Used in a Clinical Study and at the Upper and Lower Limits of the In Vitro Specification

Parameter	Reference clinical lot	Test lot lower limit	Ratio of test/reference lower limit	Test lot upper limit	Ratio of test/reference upper limit
C_{max} (ng/mL)	286 ± 61	230 ± 52	80.3	334 ± 70	116.6
T_{max} (hr)	12.9 ± 0.5	16.5 ± 0.4	127.9	10.8 ± 0.4	83.5
AUC_{inf} (ng·hr/mL)	6300 ± 1600	6000 ± 1500	94.8	6470 ± 1600	102.8

IN VITRO–IN VIVO CORRELATION FOR NONORAL DOSAGE FORMS

Current regulatory guidance on IVIVC is limited primarily to oral dosage forms. Although the discussion thus far has focused on oral dosage forms, the principles for oral dosage forms may also be applied (with modification if necessary) to novel nonoral dosage forms such as implants, drug-eluting stents, and transdermal or liposomal products. Novel delivery systems pose particular challenges in developing an IVIVC including the following:

1. The long duration of delivery. This applies particularly to drug-eluting stents, depots, implants, and multiday transdermal systems (such as three-day transdermal system for delivery of fentanyl or seven-day systems for contraceptives). Many of these delivery systems release drugs over days or months, whereas in vitro release studies are typically designed to release drug over a time frame of hours.
2. Lack of an appropriate in vitro release medium that effectively reflects the in vivo behavior.

Drug-Eluting Stents

Stents are small tubular metal scaffolds that are placed in blood vessels to maintain vessel patency. A major complication associated with bare metal stents is restenosis, related perhaps to the intrinsic thrombogenicity of the alloy or the materials used, or due to injury associated with deployment of the stent. The pathogenesis of restenosis involves proliferation and migration of smooth muscle cells from the injured artery into the stent. One of the advances in interventional cardiology is the availability of drug-eluting stents. These stents are designed to deliver drugs (generally anti-inflammatory or antineoplatic agents) locally to inhibit leukocyte infiltration and activation and proliferation of smooth muscle. It is difficult to fully apply the principles of classical IVIVC to drug-eluting stents, in part because the goal is local delivery and less so systemic delivery. Several publications have correlated the in vitro release kinetics of paclitaxel (16) and dexamethasone (17) with delivery into the artery wall with some success.

Implants

VIADUR® (leuprolide acetate implant) is a sterile, nonbiodegradable, single-use, titanium implant designed to deliver leuprolide continuously at a constant, zero-order rate over one year for the palliative treatment of advanced prostrate cancer (18). The in vivo performance of the osmotic implant systems was evaluated in Fisher rats. Systems were explanted at 3, 6, 9, and 12 months following implantation, and the residual drug in the implants is measured by reversed-phase high-performance liquid chromatography. In vitro release from the systems

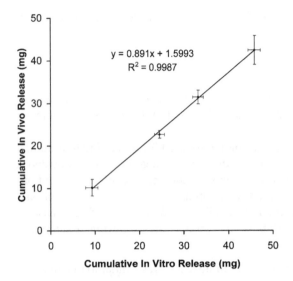

Figure 5 Comparison of the cumulative in vitro and in vivo release for an implant, demonstrating good in vitro–in vivo correlation.

was evaluated by placing the systems in test tubes containing phosphate buffered saline with 2% sodium azide as preservative. Test tubes were maintained in a water bath at 37°C. There was a very good linear correlation between the in vitro and in vivo release (Fig. 5).

Transdermal Products

Several examples showing good correlations between the in vitro release and in vivo performance of transdermal systems are available (19–22). Commonly, a convolution approach is used to demonstrate IVIVC. It is assumed that the shape of the in vivo release rate is identical to the in vitro release. Differences in the total amount released may be corrected by appropriate scaling. For a liquid reservoir transdermal system, the drug release rate is described by Fick's first law of diffusion:

$$\frac{dQ}{dt} = AD_m \frac{k_m}{h_m} C$$

where dQ/dt is the drug release rate (mass/time), A the surface area (cm^2), D_m the diffusion coefficient of drug in the membrane (cm^2/hr), k_m the partition coefficient between the membrane and vehicle, h_m the thickness of the membrane (cm), and C the concentration of drug in the vehicle assuming sink conditions (mg/cm^3). Most compounds are thought to permeate through the skin by passive diffusion, with the permeation rate given by Fick's first law of diffusion.

Assuming the skin is a homogenous single layer, the transdermal absorption rate, k_a, is as follows:

$$k_a = Ak_s \frac{D_s}{H_s V_s}$$

where k_s is the partition coefficient between the stratum corneum and the dermis, V_s the volume of skin, H_s the thickness of skin, and D_s the diffusion coefficient through the skin.

Compartmental approaches are commonly used for describing transdermal pharmacokinetics. Gupta et al. (19) used the in vitro release rate and intravenous pharmacokinetic data for fentanyl to demonstrate the use of a convolution-based IVIVC approach with good correlation between the predicted and observed serum fentanyl concentrations, following application of the transdermal therapeutic system. To establish the model, the following assumptions were made.

1. The shape of the in vitro release-rate curve was similar to that of the in vitro release (the total amount could vary).
2. Percutaneous absorption was assumed to a first order-process.
3. Since codelivery of permeation enhancers (ethanol in this case) can affect absorption, the absorption process was modified to account for the effect of ethanol.

Murthy et al. (22) demonstrated a multiple Level C correlation for a salbutamol sulfate transdermal delivery system. In vitro diffusion studies were conducted using freshly excised human cadaver epidermis from the chest region placed in a diffusion cell. In vivo data were from a pharmacokinetic study conducted in six subjects. The salbutamol transdermal system was applied onto the anterior surface of the forearm for 24 hours. A linear correlation ($R^2 = 0.99$) was noted between the cumulative amount of drug diffused in vitro and the cumulative AUC_{0-t} for various time points. This study used a single transdermal formulation and so it could not be established whether a Level A correlation was possible.

Liposomal Products

Liposomes are microvesicles composed of a bilayer of lipid amphiphetic molecules enclosing an aqueous environment. Liposomal products are formed when a drug is encapsulated within the lipid bilayer or in the interior aqueous environment. Liposomal products are designed to exhibit a different pharmacokinetic profile (typically more sustained systemic residence and different distribution) compared with the unencapsulated drug. The in vivo stability of liposomes may be affected by interactions with lipoproteins and proteins normally present in the blood. Although extensive formulation development occurs in vitro before in vivo testing, there are often instances where the in vivo performance of a liposomal formulation does not match the in vitro performance. A valid in vitro assay that provides appropriate simulation of the in vivo physiological condition would be important for assessing the quality of a liposomal product, assessing the adequacy of process controls, and providing release characteristics for the product, and may allow assessment of chemistry, manufacturing, and controls changes on the in vivo performance. Various approaches have been used to facilitate measurement of drug release in vitro (23). These methods have met with varying degrees of success in improving the concordance between in vitro and in vivo drug release. Potential reasons for this lack of correlation may relate to the vast lipid membrane pool or "sink" present in vivo to which released drug can bind. Recently, Shabbits et al. (24) presented an in vitro method using an excess of empty multilamellar vesicles that showed good correlation with the in vivo release for the model compounds, doxorubicin, verapamil, and ceramide.

USE OF PHARMACODYNAMICS IN DEVELOPING IN VITRO–IN VIVO CORRELATIONS

The discussion thus far has focussed on the use of drug concentration in relevant biofluids (blood concentration or urinary excretion) to assess systemic availability. There are also instances where a pharmacological response may be used to assess drug absorption, obviating the need to measure drug concentrations. In addition, information derived from the pharmacological response with regard to drug availability has the same utility as that derived from measurement of blood concentrations or urinary excretion data. Implicit assumptions in this approach are (*i*) that the rate limiting step in the process manifesting the pharmacological response is availability of the drug at the relevant biophase, that is, the steps associated with receptor interaction and transduction are rapid; (*ii*) all processes are reversible, that is, removal of drug from the biophase results in dissipation of the pharmacological response; and (*iii*) that the biokinetic behavior of the drug is linear (25). Methods based on pharmacological response would have greatest utility where it may not be practical or convenient to measure drug concentrations. Drug absorption analysis based on pharmacological response has been illustrated using drugs such as tropicamide (25) and warfarin (26).

CONCLUSIONS

This chapter outlines a general process of developing and applying IVIVC that works in a majority of cases. It should be recognized that successful development of an IVIVC involves exploration, and some would say it is partly an art. There are numerous cases where the principles outlined in this chapter need modification to accommodate the drug under consideration.

REFERENCES

1. Brockmeier D, Dengler HJ, Voegele D. In vitro-in vivo correlation of dissolution, a time scaling problem? Transformation of in vitro results to the in vivo situation, using theophylline as a practical example. Eur J Clin Pharmacol 1985; 28(3): 291–300.
2. Chan KK, Langenbucher F, Gibaldi M. Evaluation of in vivo drug release by numerical deconvolution using oral solution data as the weighting function. J Pharm Sci 1987; 76(6):446–450.
3. Center for Drug Evaluation and Research (CDER). Extended release solid dosage forms: development, evaluation and application of in vitro/in vivo correlations. US Department of Health and Human Services, Food and Drug Administration, Guidance for the Industry. Sep 1997.
4. The United States Pharmacopeia. In vitro and in vivo evaluation of dosage forms. < 1088 > 23rd ed. Rockville: U.S. Pharmacopeial Convention, 1995:1924–1929.
5. Uppoor RS, Marroum PJ. Regulatory considerations for oral extended release dosage forms and in vitro (dissolution)/in vivo (bioavailability) correlations. In: Sahajwalla CH, ed. New Drug Development. Regulatory paradigms for clinical pharmacology and biopharmaceutics. New York: Marcel Dekker, Inc., 2004:417–448.
6. Center for Drug Evaluation and Research (CDER). SUPAC-MR: modified release solid dosage forms. Scale-up and postapproval changes: chemistry, manufacturing, and controls; In Vitro dissolution testing and in vivo bioequivalence documentation. US Department of Health and Human Services, Food and Drug Administration. Sep 1997.
7. Dunne A, O'Hare T, Devane J. Level A in vivo-in vitro correlation: nonlinear models and statistical methodology. J Pharm Sci 1997; 86(11):1245–1249.
8. Modi NB, Lam, A, Lindemulder E, Wang B, Gupta SK. Application of an in vitro-in vivo correlation (IVIVC) in setting formulation release specifications. Biopharm Drug Dispos 2000; 21(8):321–326.
9. Pitsiu M, Sathyan G, Gupta S, Verotta S. A semiparametric deconvolution model to establish in vivo-in vitro correlation applied to OROS® oxybutynin. J Pharm Sci 2001; 90(6):702–712.
10. O'Hara T, Hayes S, Davis J, Devane J, Smart T, Dunne A. In vivo-in vitro correlation (IVIVC) modeling incorporating a convolution step. J Pharmacokin Pharmacodyn 2001; 28(3):277–298.
11. Langenbucher F. Handling of computational in vitro/in vivo correlation problems by Microsoft Excel: I. Principles and some general algorithms. Eur J Pharm Biopharm 2002; 53(1):1–7.
12. Langenbucher F. Handling of computational in vitro/in vivo correlation problems by Microsoft Excel: III. Convolution and deconvolution. Eur J Pharm Biopharm 2003; 56(3):429–437.
13. Amidon GL, Lennernas H, Shah VP, Crison JR. A theoretical basis for a biopharmaceutic drug classification: the correlation of in vitro drug product dissolution and in vivo bioavailability. Pharm Res 1995; 2(3):413–420.
14. Langenbucher F, Mysicka J. In vitro and in vivo deconvolution assessment of drug release kinetics from oxprenolol Oros preparations. Br J Clin Pharmac 1985; 19(suppl 2):151S–162S.

15. Hwang SS, Gorsline J, Louie J, Dye D, Guinta D, Hamel L. In vitro and in vivo evaluation of a once-daily controlled-release pseudoephedrine product. J Clin Pharmacol 1995; 35(3):259–267.
16. Finkelstein A, McClean D, Kar S, et al. Local drug delivery via coronary stent with programmable release pharmacokinetics. Circulation 2003; 107(5):777–784.
17. Lincoff AM, Furst JG, Ellis SG, Tuch RJ, Topol EJ. Sustained local delivery of dexamethasone by a novel intravascular eluting stent to prevent restenosis in the porcine coronary injury model. J Am Coll Cardiol 1997; 29(4):808–816.
18. Wright JC, Leonard ST, Stevenson CL, et al. An in vivo/in vitro comparison with a leuprolide osmotic implant for the treatment of prostrate cancer. J Contr Rel 2001; 75(1–2):1–10.
19. Gupta SK, Southam M, Gale R, Hwang SS. System functionality and physicochemical model of fentanyl transdermal system. J Pain Symptom Manag 1992; 7(suppl 3): S17–S26.
20. Gupta SK, Bashaw ED, Hwang SS. Pharmacokinetic and pharmacodynamic modeling of transdermal products. In: Shah VP, Maiback HI, eds. Topical Drug Bioavailability, and Penetration. New York: Plenum Press, 1993:311–332.
21. Qi X, Liu R, Sun D, Ackermann C, Hou H. Convolution method to predict drug concentration profiles of 2,3,5,6-tetramethylpyrazine following transdermal application. Int J Pharm 2003; 259(1–2):39–45.
22. Murthy SN, Hiremath SR. Clinical pharmacokinetic and pharmacodynamic evaluation of transdermal drug delivery systems of salbutamol sulfate. Int J Pharm 2004; 287(1–2):47–53.
23. Peschka R, Dennehy C, Szoka FC. A simple in vitro model to study the release kinetics of liposome encapsulated material. J Control Release 1998; 56(1–3):41–51.
24. Shabbits JA, Chiu GNC, Mayer LD. Development of an in vitro drug release assay that accurately predicts in vivo drug retention for liposomal-based drug delivery systems. J Control Release 2002; 84(3):161–170.
25. Smolen VF, Schoenwald RD. Drug-absorption analysis for pharmacological data I: method and confirmation exemplified for the mydriatic drug tropicamide. J Pharm Sci 1971; 60(1):96–103.
26. Smolen VF, Erb RJ. Predictive conversion of in vivo drug dissolution data into in vivo drug response versus time profiles exemplified for plasma levels of warfarin. J Pharm Sci 1977; 66(3):297–304.

6

IVIVC for Oral Drug Delivery: Immediate Release and Extended Release Dosage Forms

Colm Farrell and Siobhan Hayes

*Department of Pharmacokinetics and Biopharmaceutics,
GloboMax—ICON Development Solutions, Marlow, Bucks, U.K.*

INTRODUCTION

The correlation between the in vitro dissolution performance of an oral dosage form, particularly those employing proprietary delivery technologies, to the corresponding in vivo performance has long been an important goal for the pharmaceutical scientist. However, in more recent times, interest in the field of in vitro–in vivo correlations (IVIVC) has grown, reflecting the potential of this methodology as a tool for optimizing formulation development as well as a tool for supporting applications to regulatory authorities.

This chapter will focus primarily on the development, validation, and application of IVIVC models for oral dosage forms for both regulatory applications and for formulation development.

Almost a decade has passed since the publication of the 1997 FDA Guidance on IVIVC for oral modified release (MR) products. In this time, both companies and the regulatory agencies have gained valuable experience in the development, validation, and application of IVIVC models. As with other disciplines, "best practise" has evolved since the IVIVC guidance was first published, although many of these practises are not formally documented. This chapter will also endeavor to highlight the current state of play regarding IVIVC and oral drug delivery.

IVIVC AND EXTENDED RELEASE DOSAGE FORMS

The concept of IVIVC for extended release (ER) dosage forms that allows scientists to predict the expected bioavailability for an ER product based on its in vitro dissolution profile has long been discussed by the pharmaceutical community. Over the years, a number of seminal workshops and publications have tracked the progress made in this area. A 1987 ASCPT/DIA/APS/FDA-sponsored workshop concluded that the available science and technology at that time did not permit consistently meaningful IVIVC for ER dosage forms and encouraged IVIVC as a future objective (1). The workshop also concluded that in vitro dissolution testing could only be considered useful for stability, process control, minor changes to formulations, and changes in manufacturing site. The following year, 1988, saw the publication of a USP PF stimuli article in which the classification of IVIVC models into Levels A, B, and C was established — a classification that is still in use today (2). A report from an ASCPT/DIA/APS/FDA-sponsored workshop in 1990 concluded that, although the science and technology available did not always allow for meaningful IVIVC, the development of such models was considered an important objective on a product-by-product basis, and procedures for the development, evaluation, and application of an IVIVC were described (3). The report also included the suggestion that proposed in vitro dissolution specifications should be validated by conducting a bioequivalence study involving two batches with in vitro dissolution profiles representative of the upper and lower specifications. Another ASCPT/DIA/APS/FDA-sponsored workshop in 1993 resulted in a report that provided further information related to IVIVC (4). The objectives of an IVIVC were identified as being the use of in vitro dissolution testing as a surrogate for bioequivalence (BE) testing, as well as being an aid in the setting of dissolution specifications. This report concluded that in vitro dissolution might be used as a sensitive, reliable, and reproducible surrogate for bioequivalence testing. The report also supported the concepts contained in USP chapter 1088 (5), which described appropriate techniques for Level A, B, and C correlations and methods for establishing dissolution specifications. The report from the 1993 workshop also found that IVIVC could be useful for changes other than minor changes in formulation, equipment, process, manufacturing site, and batch size.

Quite clearly, the 1993 workshop report reflected the increased confidence within the pharmaceutical industry at that time in the use of IVIVC to estimate the in vivo bioavailability characteristics of an ER drug product and the willingness to rely on in vitro dissolution testing and IVIVC to serve as a surrogate for bioequivalence studies. All this is in marked contrast to the conclusions only six years earlier of the 1987 workshop.

The heightened awareness of IVIVC and its potential application to both the formulation development process and the support of regulatory submissions for ER dosage forms culminated with the publication in 1997 of the FDA Guidance for Industry—ER oral dosage forms: development, evaluation, and application of IVIVCs (6). This guidance describes the levels of correlations that can be established, but more critically also established the varying degrees of usefulness of the different levels of IVIVC. The guidance also provides information on issues related to study design and conduct, how IVIVC models are to be evaluated, and the practical applications of IVIVC.

The 1997 FDA guidance on IVIVC and oral ER dosage forms still serves today as the only specific guidance on IVIVC. However, the principles and applications laid down in this document should serve those interested in developing IVIVC for alternate dosage forms, particularly those that utilise controlled release technology.

IVIVC AND IMMEDIATE RELEASE DOSAGE FORMS

Historically, attempts to successfully develop IVIVC models for oral immediate release (IR) dosage forms have been less successful than for controlled release or sustained release dosage forms, particularly for Level A IVIVC models. A number of factors contributed to the lack of success: in vitro dissolution evaluation of IR dosage forms frequently only included one to three time-points, making the development of Level A models almost impossible; many of the IR dosage forms did not exhibit dissolution-rate limited absorption; the relationship between in vitro dissolution and in vivo absorption frequently appeared to be nonlinear as in vivo absorption could not "keep up" with in vitro dissolution. This last factor did not preclude the development of Level A models but at the time, Level A models were defined as "point-to-point" relationships and were anticipated to be linear. As a consequence, IVIVC was not attempted for many IR dosage forms because of the low expectation for success.

However, our understanding of this apparent lack of success with IVIVC for IR dosage forms has increased with the advent of the biopharmaceutical classification system (BCS) (7). As discussed elsewhere in this compilation, the BCS, as defined by Amidon et al. provides a framework for classifying drug substances based on their aqueous solubility and intestinal permeability. The BCS takes into account the three major factors that govern rate and extent of absorption from IR solid oral dosage forms: dissolution, solubility, and intestinal permeability. On this basis, drug substances are classified into four classes: (*i*) class I for substances with both high solubility and permeability, (*ii*) class II for substances with low solubility and high permeability, (*iii*) class III for substances with high solubility and low permeability, and (*iv*) class IV for substances with both low solubility and permeability.

For drug substances in class I, an IVIVC is anticipated if dissolution is slower (i.e., rate controlling) compared to gastric emptying. Otherwise, a limited or no correlation is expected for rapidly dissolving formulations of a class I drug substance. For class II drug substances, an IVIVC is expected if in vitro dissolution is similar to in vivo dissolution rate. For drug substances in class III, absorption (permeability) is the rate-controlling factor, not dissolution.

Therefore, only a limited or no correlation between in vivo absorption and in vitro dissolution rate is anticipated. Finally, no correlation is expected for class IV drug substances unless in vivo dissolution is rate controlling.

The BCS has helped us understand some of the historical lack of successful IVIVC attempts for IR dosage forms and also helps us to anticipate the likelihood of success for future drug substances, based on their classification.

REGULATORY FRAMEWORK FOR IVIVC AND ORAL DOSAGE FORMS

Although the 1997 FDA guidance is the only specific IVIVC guidance to be published to date, for any pharmaceutical scientist with an interest in the area of IVIVC, there are a number of additional guidances that are associated with this area that provide the framework for the regulatory application of IVIVC. These include the guidances on scale-up and post approval changes (SUPAC) for both modified release and IR solid oral dosage forms (8,9). There is also guidance on the waiver of in vivo studies for IR solid oral dosage forms based on the biopharmaceutics classification of the drug (10). The recent guidance on bioavailability and bioequivalence studies for oral products also provides information on the application of IVIVC models (11).

The Committee for Proprietary Medicinal Products (CPMP) within the European Agency for the Evaluation of Medicinal Products (EMEA) has also issued a note for guidance on the pharmacokinetic and clinical evaluation of modified release oral products that provides some information on the development and evaluation of an IVIVC (12).

A review of these guidances should quickly inform the scientist that the area of IVIVC requires the collaboration of formulators, analytical chemists, pharmacokinetists and regulatory personnel to provide an integrated approach to the development and application of IVIVC models in both the formulation development and regulatory submission process.

At present, no specific guidance has been issued dealing with IVIVC and oral IR dosage forms. However, the applications of IVIVC for oral IR dosage forms are discussed in both the FDA Guidance for Industry on dissolution testing of IR dosage forms (13) and in the SUPAC-IR guidance.

DEVELOPMENT AND VALIDATION OF LEVEL A IVIVC MODELS FOR ORAL DOSAGE FORMS

Before discussing the application of IVIVC models to the development of oral dosage forms, it is worth spending a few moments briefly reviewing the development and validation of such models.

The 1997 FDA IVIVC guidance for extended-release oral products defines an IVIVC as "a predictive mathematical model" that describes the relationship between an in vitro property of an ER dosage form (usually the rate or extent of drug dissolution or release) and a relevant in vivo response (e.g., plasma drug concentrations or the amount of drug absorbed). Originating from the USP stimuli article (2), IVIVC models have been classified into different main categories: Levels A, B, and C. The category of IVIVC model depends entirely on the analysis performed—in all cases, the same in vitro (dissolution-time data) and in vivo data (plasma concentration-time data) are collected. Critically, the usefulness of an IVIVC model, in terms of both internal decision making during drug development programs and for regulatory submissions, is determined by the category of model.

Level C models are simple single-point relationships between the amount dissolved in vitro at a particular time or the time required for in vitro dissolution of a fixed percentage of the dose, for example, $t_{50\%}$ and a summary parameter that characterizes the in vivo time course, for example, C_{max} or AUC. By its very definition, Level C IVIVC models need at least three formulations and it is important to note that a Level C correlation does not reflect the complete shape of the plasma concentration time curve.

Multiple Level C models relate one or several pharmacokinetic parameters of interest to the amount of drug dissolved at several time points along the dissolution profile, such as the amount released at certain time points or the time to release a certain percentage, that is, $t_{50\%}$. Multiple Level C models can be useful in identifying critical dissolution time points to assist in setting dissolution specifications. According to FDA Guidance, multiple Level C models can be as useful as Level A models and both the the FDA and CPMP suggest that if multiple Level C models can be developed, a Level A IVIVC is also likely. However, our experiences suggest that multiple Level C correlations may not be acceptable to all regulatory authorities.

Level B IVIVC models use the principle of statistical moment analysis. For example, the mean in vitro dissolution time could be compared to the mean residence time or mean in vivo dissolution time. Such models are considered to be the least useful for regulatory purposes and will not be discussed further.

Finally, Level A IVIVC models describe the relationship between the entire in vitro dissolution time course and the entire in vivo response time course for example, the time course of plasma-drug concentration or amount of drug absorbed. From a regulatory viewpoint, these models are considered to be the most informative and are recommended. Therefore, the focus of this section will be on the development and evaluation of Level A IVIVC models for oral dosage forms.

The development and evaluation of a Level A IVIVC model can be broken down into a series of steps:

1. Selection of appropriate formulations
2. Obtaining in vitro dissolution data for these formulations
3. Design and conduct of the in vivo study itself
4. Data analysis

The likelihood of obtaining a successful outcome to the development of an IVIVC model can be increased by considering the first three topics listed above. The reader is directed to a comprehensive review by Shepard et al. (14) of the issues related to study design considerations for IVIVC studies. Therefore, the present chapter will confine itself to a brief overview of the development and validation of Level A IVIVC models to facilitate the later discussions regarding the applications of such models.

Level A IVIVC models have historically been developed using a two-stage approach. The first stage involves deriving the in vivo release profile from the observed plasma concentration data using a suitable deconvolution method and the second stage involves the modelling of the relationship between the derived in vivo release and the in vitro release data.

Since the release of the 1997 IVIVC guidance, experience gained from both regulatory submissions and subsequent discussions at scientific meetings, a number of key issues regarding the two-stage approach have emerged: firstly, it is expected that the deconvolution procedure takes place on an individual subject basis. That is, the in vivo release profile should be obtained for each individual, rather than averaging the concentration-time data before performing the deconvolution procedure. Following the individual deconvolution, the average in vivo input function is obtained and it is this entity that is subsequently correlated with the average in vitro dissolution data to obtain an IVIVC model. The output of the IVIVC model itself is a single predicted in vivo input function for each formulation.

In order to validate the model, the output from the IVIVC model needs to be translated into concentration-time profiles for each treatment and this is done by performing the reverse operation of deconvolution, namely convolution. The result of the convolution procedure is a single predicted concentration-time profile for each formulation from which summary parameters AUC and C_{max} are obtained and compared to the corresponding parameters derived from the average observed data.

For internal predictability, the IVIVC model is used to predict the AUC and C_{max} for each of the formulations or batches used in the development of the model. To satisfy the FDA criteria, the AUC and C_{max} absolute percent prediction error (%PE) should be no greater than 15% for any single formulation and the mean absolute percent prediction error (MAPPE) should be less than 10% for each parameter. Where the MAPPE falls within 10% to 20%, the IVIVC model is considered to be inconclusive and external predictability should be conducted.

To demonstrate external predictability, the IVIVC model is used to predict the AUC and C_{max} of a formulation or batch that was not included in the original model development. This could be a formulation that was evaluated in the same study as those formulations used in the model development or it could be a formulation evaluated in a separate study to that used to develop the IVIVC. The criteria for external predictability are that the %PE should be less than 10%. Therefore, these criteria may be perceived as being more stringent that those for internal predictability, in that a formulation can have a %PE of up to 15% for internal predictability but must be less than 10% for external predictability.

Although the criteria for IVIVC model validation discusses AUC and C_{max} only, our recent experiences indicate that regulatory agencies also place emphasis on how well an IVIVC model predicts the entire plasma concentration time course, for example, how well does the model predict the time at which the maximum concentration is observed and how well the IVIVC model describes the terminal phase of the data.

APPLICATION OF IVIVC—BASIC PRINCIPLE

The various applications of IVIVC for oral dosage forms are all based on the same principle—that is, a validated IVIVC model allows us to predict the impact of changes to a formulation on the in vivo performance without having to conduct a BE study.

In the absence of a validated IVIVC model, the management of change to a dosage form is typically done according to the following procedure. For example, the new manufacturing process or revised formulation is used to produce the new dosage form, which is subjected to in vitro dissolution testing. Based on the results, a decision is then taken whether to proceed with a standard two-way, crossover BE study between batches produced with the new and the existing process/formulation. If the two products are demonstrated to be bioequivalent, then the new process/formulation is substituted for the existing one in the development program. However, if BE cannot be demonstrated, the cycle starts all over again. More costly though are the implications of the bioinequivalence on the drug development program, being more severe the further down the development program the change is made.

Having a validated IVIVC, the process is similar, but now the decision regarding BE is taken on the basis of the results of the in vitro dissolution test and the IVIVC model, by predicting concentration-time profiles for the batches produced with the new and the existing process/formulation and determining the differences.

Thus, although the most obvious saving between the two approaches may be financial in terms of the money saved in not performing BE studies, the bigger saving is in time. An IVIVC model allows the impact of a change on the in vivo performance to be determined in a matter of minutes or hours, rather than in months. In the current drug development climate, this is critically important as it avoids decisions being taken at risk, pending the results of a BE study. Thus, the value of having an IVIVC in the oral dosage form development program is that it allows more timely and reliable decisions to be taken regarding the potential impact of changes on the in vivo performance of the dosage form.

One critical point regarding the application of IVIVC models is that the predicted concentration-time profile and corresponding parameters for the new process/formulation should not be compared to historical observed data for the existing process/formulation. The rationale for this is quite simple: any IVIVC model has an associated error in terms of its ability to predict data. For validated IVIVC models, the error has been quantified and demonstrated to satisfy the criteria outlined in the guidance and discussed earlier in this chapter. For example, a validated IVIVC model predicts AUC for the target formulation with an associated prediction error of 10%. A second site manufactures a batch of the same target formulation and in vitro dissolution testing shows it has an identical profile to the formulation manufactured at the current site. However, comparison of the IVIVC model-predicted AUC for the batch from the new site and the observed AUC for the batch from the current site reveals an apparent 10% differ-ence. So, for identical in vitro dissolution profiles, a 10% difference in AUC is apparently predicted. Clearly, this represents the already quantified prediction error associated with the IVIVC model. Using the IVIVC model to predict the in vivo performance for both batches would show no difference in AUC, as anticipated from identical in vitro release profiles. This approach is also consist-ent with the notion that an IVIVC model acts as a surrogate for a BE study. Were such a study to be performed, a batch from each of the current and new manufac-turing sites would be evaluated in the same study.

REGULATORY APPLICATION OF IVIVC FOR IMMEDIATE RELEASE DOSAGE FORMS

As previously discussed, successful development of IVIVC models for IR dosage forms may be limited to those drug substances in either class II or III, according to the BCS. Despite this apparent limitation, IVIVC, where developed, can be applied for IR dosage forms for regulatory purposes.

In the 1995 SUPAC-IR Guidance, the use of an "acceptable IVIVC" is discussed to support the waiving of a full BE study where a change requires in the in vivo BE documentation. Such documentation is only required for a Level 3 change in components and composition or a Level 3 change in the manufactur-ing process (i.e., change to a different campus).

IVIVC models can also be applied to the development of in vitro dissol-ution specifications for IR dosage forms.

REGULATORY APPLICATION OF IVIVC FOR EXTENDED RELEASE DOSAGE FORMS

The IVIVC guidance for ER oral dosage forms discusses in some detail the appli-cations of an IVIVC. The guidance does stress, however for these applications, that predictability of the IVIVC model must have been established if it is to serve as a surrogate for in vivo testing. The IVIVC guidance refers the reader to the SUPAC-MR guidance for industry, where the requirement for BE documentation for various levels of scale-up and post-approval changes. Where such documentation is required, the need to conduct a formal study may be waived in the presence of an established IVIVC. Presently, under SUPAC-MR, an IVIVC may be used to support a waiver for the following changes: (*i*) a Level 3 change in a nonrelease controlling excipient, (*ii*) a Level 2 change in release controlling excipient for a drug with a narrow therapeutic index, (*iii*) a Level 3 change in release controlling excipient, (*iv*) a Level 3 change in manufacturing site, and (*v*) a Level 3 change in manufacturing process.

The FDA IVIVC guidance also discusses how an IVIVC can be used to justify a biowaiver request for the approval of lower strengths or new strengths. In these cases, the IVIVC should have been developed using the highest strength and the new strengths should be compositionally proportional or qualitatively the same, have the same release mechanism, and have similar in vitro dissolution profiles.

Importantly, the IVIVC guidance from the FDA also qualifies the application of IVIVC models as a surrogate for an in vivo study, based on the number of formulations/release rates used in the development of the IVIVC and whether the drug has a narrow therapeutic index. For example, an IVIVC developed with only two formulations/release rates is more limited in its applications than an IVIVC developed with three or more formulations/release rates. Similarly, the demonstration of external predictability strengthens the case of applying IVIVC as a surrogate for a BE study, particularly where the model has been developed with just two formulations/release rates or where the drug has a narrow therapeutic index. Clearly, the greater the confidence in an IVIVC, the wider it can be applied in a regulatory setting.

As well as the possibility of using an established IVIVC to waive the requirement to conduct a BE study, the SUPAC-MR Guidance for Industry also discusses how an IVIVC can reduce the amount of in vitro dissolution documentation required to support Level 2 changes to a formulation or process. For all MR oral dosage forms, in the presence of an established IVIVC, only application or compendial dissolution testing should be performed (i.e., only in vitro release data by the correlating method needs to be submitted).

The FDA IVIVC Guidance for Industry outlines how a Level A IVIVC can be applied for the setting of dissolution specifications. Using the IVIVC model, predicted C_{max} and AUC are obtained for the fastest and slowest release rates that are allowed by the dissolution specifications such that a maximal difference of 20% exists. Ensuring a maximal difference of 20% in the predicted C_{max} and AUC for the upper and lower specifications is considered the optimal situation. The guidance suggests that an established IVIVC may permit the setting of wider specifications, dependent on the predictions of the IVIVC (i.e., as long as the differences in the predicted C_{max} and AUC for the lower and upper limits do not exceed 20%). This was re-affirmed in a presentation at a recent advisory committee for pharmaceutical science meeting (15).

The 1997 guidance did suggest a less optimal, but still acceptable, situation where the specifications ensure lots exhibiting dissolution profiles at the upper and lower limits are bioequivalent to the clinical/bioavailability lots or to an appropriate reference standard. The implication was that the difference in the predicted C_{max} and AUC between the lower and upper specifications could be greater than 20% as long as both were bioequivalent to the target. However, our recent experience suggests that proposed dissolution specifications that fail to ensure a maximal difference of 20% in the predicted pharmacokinetics for the upper and lower specification limits are not currently acceptable to the FDA.

The advisory committee for pharmaceutical science meeting in May 2005 also suggested that external validation is not required for the application of an IVIVC model in setting dissolution specifications. However, our recent discussions with the Agency suggest this only applies to models that are considered to be straightforward. More complex IVIVC models, especially those developed to account for dissolution rate-dependent changes in the in vivo performance of an ER product, may require evidence of external validation before the model can be applied to dissolution specifications.

APPLICATION OF IVIVC FOR DEVELOPMENT OF IN VITRO DISSOLUTION SPECIFICATIONS—A CASE STUDY

This case study briefly illustrates how a validated Level A IVIVC model for an ER dosage form was applied for the development of in vitro dissolution specifications. The IVIVC model was developed using in vitro dissolution data sampled at eight time points over 20 hours. The established model was firstly used to identify which time point(s) was critical for an accurate prediction of C_{max} for the target formulation around which the specifications were to be established. By omitting each time point in turn, predicting C_{max}, and comparing the predicted value to the value obtained using the full dissolution profile, the IVIVC model quickly indicated that only the six-hour time point was critical for the prediction of C_{max}. The IVIVC model was used to demonstrate that the concentration-time profile for the target formulations could be described using only four time points (Fig. 1A). The predicted profiles and corresponding PK parameters were then obtained for the lower and upper specifications by convolution through the established IVIVC model (Fig. 1B).

APPLICATION OF IVIVC FOR THE JUSTIFICATION OF A LEVEL 3 SITE CHANGE—A CASE STUDY

The following case study briefly described how an established IVIVC model was used to justify a Level 3 site change in manufacturing site, a change that would require documentation of BE according to SUPAC-MR, in the absence of an IVIVC model. The target formulation was manufactured at the proposed commercial manufacturing site and the in vitro dissolution profile compared to a

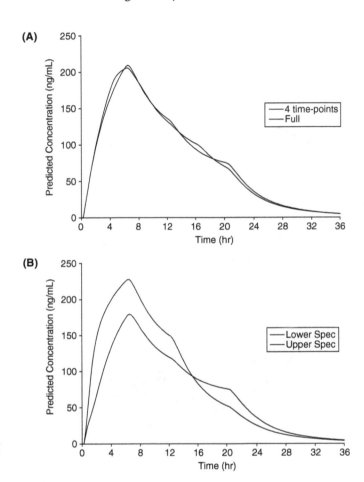

Figure 1 (**A**) Predicted concentration-time profile for the target formulation using either the full dissolution profile (eight time points) or just four key time points identified using an in vitro–in vivo correlations model. (**B**) Predicted concentration-time profiles for the lower and upper dissolution specifications. C_{max} and AUC for the lower specification were within 20% of the corresponding values for the upper specification.

batch of the identical formulation manufactured by the clinical trial supply site (Fig. 2A). The concentration-time profile and corresponding PK parameters for each batch were predicted using the established IVIVC model and shown to be bioequivalent, that is, both AUC and C_{max} within 20% (Fig. 2B).

Both case studies clearly illustrate the application of an established IVIVC model, allowing in vitro dissolution testing to serve as a surrogate for BE studies so that the impact of changes in a formulation or changes in site of manufacture can be assessed quickly and allow the development program to continue.

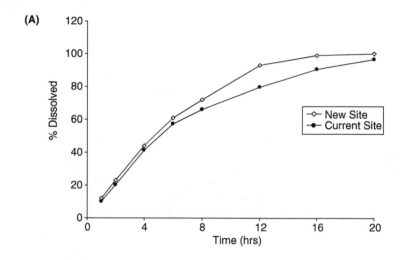

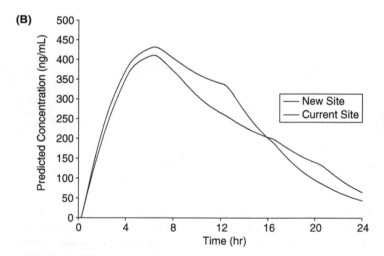

Figure 2 (**A**) In vitro dissolution profiles for a 50 mg ER formulation manufactured at the current and proposed new sites. (**B**) Concentration-time profiles predicted from the in vitro dissolution profiles using an in vitro–in vivo correlations model. C_{max} and AUC for the new site were within 5% for the corresponding values for the current site.

In the same way that postapproval changes can be supported using an IVIVC, changes made during the preapproval process may also be justified using in vitro dissolution data and an established IVIVC. However, companies need to consider the potential risk of utilizing an IVIVC that has yet to receive approval from a regulatory agency to support such pre-approval changes. Clearly, the more robust the IVIVC model, the greater confidence the company can have in its application in the pre-approval environment. For example, was the model developed using three or more different formulations/release rates? Has external predictability, in addition to internal predictability, been demonstrated for the model?

Frequently, companies now present IVIVC models at end of phase II-type meetings with regulatory agencies, and outline how they intend to incorporate such models into their BE strategy, for example, bridging the clinical trial material and the final to-be-marketed formulation. While these meetings do not provide a definitive or binding opinion from the agencies, sponsors do obtain valuable feedback as to the likely acceptance of basing a BE strategy on a particular IVIVC, and for this reason, introducing IVIVC and the intended applications at such meetings is highly recommended.

TIMING OF AN IVIVC WITHIN A FORMULATION DEVELOPMENT PROGRAM

Not surprisingly, the application of an IVIVC to any oral dosage form development is dependent on when, during the program, the IVIVC study or studies are conducted to allow a model to be developed. The timing of such studies frequently reflects the prevailing view of IVIVC within a company. For those companies that view IVIVC as being a powerful tool to assist the formulation development, a study designed to permit the development of an IVIVC model may be conducted very early in the formulation development program. A range of formulations considered sufficiently wide to allow a meaningful IVIVC model to be developed are evaluated in vivo. Typically, one of the formulations will represent what is believed to meet the product specification and formulations having either faster or slower release rates are also included in the study. If one of the formulations does meet the product specifications, the IVIVC can be used to assist the further optimization and scale-up of that formulation. If none of the formulations meets the specifications, the IVIVC can be used to guide the formulation developments and help identify formulations that more closely match the product specification. Incorporating the IVIVC model development early into the formulation program allows the model to be refined and expanded as further data become available.

Other companies are less aggressive in the application of IVIVC and tend to view IVIVC as a tool for supporting changes to an already identified target formulation. Developing an IVIVC late in the formulation development program usually requires the manufacture of formulations designed to be sufficiently faster or slower than the target to be evaluated in a prospectively designed study to allow for a meaningful IVIVC to be developed.

However, a review of the formulation development program may help identify a suitably wide range of formulations that have been evaluated over a series of early develop studies that could then be combined into a single IVIVC. This retrospective approach to IVIVC does require a common reference to be included in all early studies to allow for deconvolution to be performed. For example, early studies evaluating ER dosage forms would need to include an IR reference in each study in order to develop an IVIVC using the traditional deconvolution-based approach. Therefore, although IVIVC may not be a primary or even secondary objective of early formulation development studies, attention should be paid to the design of such studies that may permit the use of data obtained from these studies to be used for IVIVC analyses later in the development program.

Rohrs et al. approach the timing of IVIVC within an ER formulation development program from the viewpoint of the in vitro dissolution methodology (16). They suggest obtaining in vivo data on ER formulations as early as possible in the development program as any changes in the dissolution methodology are easier to implement early on the drug development before a large historical and stability database has been generated. They also recommend that samples from early formulations should be retained for any future in vitro methodology development.

SUMMARY

The role of IVIVC in the development of oral dosage forms has increased in recent years, reflecting the utility of such models, particularly Level A IVIVC, for both assisting formulation development, for setting of "biorelevant" dissolution specifications and for supporting pre- and postapproval changes that otherwise would require documentation of BE. The probability of successful IVIVC development and application within an oral development program can be considerably increased by incorporating the IVIVC strategy into the early phases of the development.

REFERENCES

1. Skelly JP, et al. Report of the workshop on CR dosage forms: issues and controversies. Pharm Res 1987; 4(1):75–78.
2. United States Pharmacopeial Convention Inc. In vitro–in vivo correlation for extended release oral dosage forms. Pharmacopeial Forum Stimuli Article 1988; July:4160–4161.
3. Skelly JP, Amidon GL, Barr WH, Lenet LZ, Carter JE, Robinson JR, et al. Report of the workshop on in vitro and in vivo testing and correlation for oral controlled/modified-release dosage forms. J Pharm Sci 1990; 79(9):849–854.
4. Skelly JP, Van Buskirk GA, Arbit HM, Amidon GL, Augsburger L, Barr WH, et al. Workshop II report scale-up of oral extended release dosage forms. Pharm Res 1993; 10(12):1800–1805.
5. United States Pharmacopeial Convention Inc. In vitro in vivo evaluation of dosage forms. USP XXIII 1927–1929.
6. Food and Drug Administration Guidance for Industry. Extended release oral dosage forms: development, evaluation, and application of in vitro/in vivo correlations, Sep 1997.

7. Amidon GL, Lennernas H, Shah VP, Crison JR. A theoretical basis for a biopharmaceutic drug classification: the correlation of in vitro drug product dissolution and in vivo bioavailability. Pharm Res 1995; 12(3):413–420.
8. Food and Drug Administration Guidance for Industry. SUPAC-MR: modified release solid oral dosage forms. Scale-up and postapproval changes: chemistry, manufacturing, and controls; in vitro dissolution testing and in vivo bioequivalence documentation, Oct 1997.
9. Food and Drug Administration Guidance for Industry. SUPAC-IR: immediate-release solid oral dosage forms: scale-up and post-approval changes: chemistry, manufacturing, and controls; In vitro dissolution testing and in vivo bioequivalence documentation, Nov 1995.
10. Food and Drug Administration Guidance for Industry. Waiver of in vivo bioavailability and bioequivalence studies for immediate-release solid oral dosage forms based on a biopharmaceutics classification system.
11. Food and Drug Administration Guidance for Industry. Bioavailability and bioequivalence studies for orally administered drug products—general considerations, Mar 2003.
12. Committee for Proprietary Medicinal Products (CPMP). Note for guidance on quality of modified release products: A. oral dosage forms; and B. transdermal dosage forms; section I (quality), July 1999.
13. Food and Drug Administration Guidance for Industry. Dissolution testing of immediate release solid oral dosage forms, Aug 1997.
14. Shepard T, Farrell C, Rochdi M. Study design considerations for IVIVC studies. In: Pharmaceutical Dissolution Testing, eds. Jennifer Dressman and Johannes Krämer New York: Marcel Dekker Inc 2006.
15. Transcript of Advisory Committee Pharmaceutical Sciences Meeting, May 3–4, 2005. (www.fda.gov/ohrms/dockets/ac/cder05.html)
16. Rohrs BR, Skoug JW, Halstead GW. Dissolution assay development for in vitro–in vivo correlations: theory and case studies. Adv Exp Med Biol 1997; 423:17–30.

7

In Vitro–In Vivo Correlation for Modified Release Parenteral Drug Delivery Systems

David Young

AGI Therapeutics, Inc., Columbia, Maryland, U.S.A.

INTRODUCTION

In the in vitro–in vivo correlation (IVIVC) breakout session at the AAPS/FDA workshop on modified release (MR) parenterals in 2001, FDA reviewers stated that, in their personal opinion, the IVIVC of parenterals may be harder to develop but the process of developing the IVIVC and the process of demonstrating their validity had to be the same as MR oral products (1,2). The rationale was based on the fact that the IVIVC was going to be used for regulatory purposes such as a biowaiver and the development of release specifications. Over the last four to five years, the general belief remains the same—the development of an IVIVC for MR parenterals should follow the same process as a MR oral product. Although the design of IVIVC studies and the IVIVC modeling process have evolved, the fundamental principles of developing and validating an IVIVC has not significantly changed regardless of the type of delivery system.

The purpose of this chapter is not to review the general principles of IVIVC, but instead, this chapter will address some of the special considerations that are required when developing an IVIVC for a controlled release parenteral product. Although the FDA guidance on IVIVC emphasizes the usefulness of Level A and multiple Level C IVIVC, this chapter will mainly address the development of a Level A IVIVC.

SPECIAL CONSIDERATIONS

Study Design

The study design for MR parenteral drug delivery systems should be similar to the design for oral formulations, if logistically possible. Typically, two or more formulations with different in vitro and in vivo release rates and formulation characteristics are administered to normal human volunteers. Patients may be used in the study if administration of the drug to normal volunteers is unsafe, which is more of a concern when administering a parenteral product, or the patient population significantly handles the drug and/or delivery system differently. The active drug in solution (defined here as the reference formulation) is also administered intravenously or through the same route of administration as the MR formulation in order to perform a Level A IVIVC. Although a complete crossover study is preferred, the logistical problems associated with running such a study may make the study impractical, given the time-course of in vivo delivery (e.g., implant delivery over a number of months). If the complete crossover study design is not possible, incomplete block and parallel designs have also been used. Regardless of the design, every subject should receive the reference formulation as the first arm of the study in order to define the unit impulse response and to ensure that a deconvolution or estimation of in vivo release can be performed, even if a subject drops out after receiving only one of three MR formulations.

In Vitro Release System

Although in vitro release (IVR) systems are well established for all types of oral formulations, standard IVR systems for MR parenterals do not exist. The literature reports a range of systems from the destructive test tube to the USP 4 or USP 7 apparatus. Although the IVR system is critical to the IVIVC modeling, this chapter will concentrate only on the modeling aspects.

In Vivo Release

With oral or any other systemic drug delivery system the in vivo release is typically estimated using the measurement of drug within the systemic circulation (i.e., studying the pharmacokinetics). However, with some parenteral products the delivery system is designed to administer drug locally. In this situation, the systemic measurement of drug does not represent what is occurring locally, or it may be difficult to obtain systemic measurements of drug because the concentration in the blood is below the limit of quantitation. For these local drug delivery systems, determining the amount of drug released over time requires invasive, destructive measurements of the delivery system to determine how much drug remains in the delivery system over time. This often cannot be determined in humans so the in vivo release must be determined in animals for the IVIVC (e.g., drug eluting stents). The FDA has been amenable to this approach, given the impossible task of measuring in vivo release in humans.

In Vitro–In Vivo Correlation Using More Complicated Modeling Approaches

For all parenteral IVIVC models, the author strongly recommends that at least three different formulations be studied in order to perform both internal and external predictability. In addition, with some MR parenteral dosage forms, in vitro release occurs over hours or days while complete in vivo release may take days, weeks, or months. The linear IVIVC models developed in the 1970s and 1980s could not deal with this time difference between the two releases. Over the last decade, time variant models (1,3,4) have been introduced and used to deal with the differences in the time course of release. A model that has provided an enormous amount of flexibility in its ability to fit time variant and linear time invariant IVIVC data has been the model described by Gillespie (3) and others (5,6). Both time shifting and time scaling can be described by the model, which allows the model to fit a wide variety of in vitro-in vivo profiles. The model used to describe both time shifting and scaling is presented in Equation 1:

$$x_{\text{vivo}}(t) = \begin{cases} 0 & t < 0 \\ & u = t \text{ for } t \leq T \\ a_1 + a_2 \cdot x_{\text{vitro}}(-b_1 + b_2 \cdot u) & u = T \text{ for } t > T \end{cases} \tag{1}$$

where $x_{\text{vivo}}(t) = $ cumulative amount absorbed or released in vivo, $x_{\text{vitro}} = $ cumulative amount released in vitro, $a_1 = $ intercept for a linear IVIVC, $a_2 = $ slope for a linear IVIVC, $b_1 = $ coefficient representing a time shift between in vivo and in vitro, and $b_2 = $ coefficient representing a time scaling between in vitro and in vivo. If $b_1 = -1$ and $b_2 = 0$, the IVIVC is the linear "point-to-point" model that has been reported in the literature over the years.

Predictable models have been developed using this approach for oral and parenteral delivery systems. An example of this model is illustrated in Figure 1. The in vivo versus in vitro release of four formulations are presented in Figure 1. Two of the formulations (K1, K2) have faster in vivo release than in vitro, while two of the formulations have faster in vitro release (K3, K4). Since an IVIVC now requires that one model be developed for two or more formulations, it would be impossible to develop one model to describe all four formulations using a conventional linear time invariant model. However, using Equation 1 to describe the shift and scaling, b_1 and b_2 can be estimated to obtain a single time variant model for all four formulations. The %PE of C_{max} and AUC for each formulation met the FDA criteria for internal predictability.

EXAMPLES

Microspheres

Typically, there are two types of profiles that have been seen with MR microsphere delivery systems: Type 1 with one peak and continuous delivery (Fig. 2, Product A) and Type 2 with an initial burst peak and a second peak at

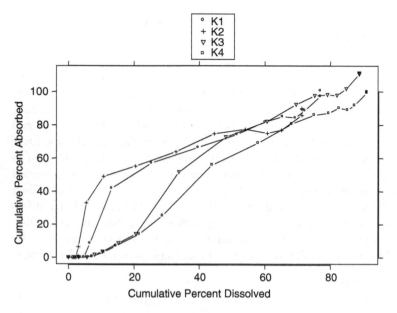

Figure 1 Cumulative percent absorbed versus cumulative percent dissolved with no time shifting.

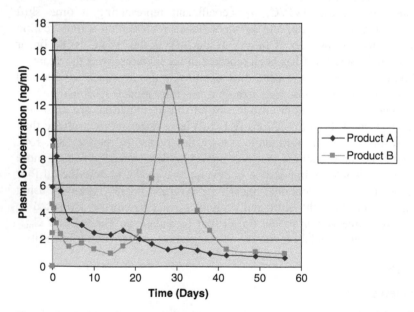

Figure 2 Plot of plasma concentration versus time after two different microsphere products were administered, Product A with an early peak and prolonged continued release and Product B with an initial release and a second release.

a later time (Fig. 2, Product B). The ideal approach to IVIVC modeling is to develop one IVIVC model for the total plasma profile. There are a number of reports on the in vitro–in vivo relationship for a single Type 1 formulation, but an IVIVC requires two ore more formulations. IVIVC models with two or more formulations have been developed for these Type 1 formulations but have not been reported in the literature (D. Young, personal communication, 2002). One of the few examples of investigating multiple formulations was reported by Van Dijkhuzien-Radersma et al. (7). These investigators reported the in vitro and in vivo relationship for two formulations but did not develop a mathematical model for both formulations. Figure 3 shows that one Level A IVIVC model could have been developed and probably would have met the internal predictability criteria.

Investigators have attempted to develop a single IVIVC model for Type 2 plasma profiles using the two-stage deconvolution and the compartmental modeling approaches. In order to develop a single model, the IVR system must be able to correlate to the very fast absorption rate of the first plasma peak and the slower absorption rate of the second plasma peak. Figure 4 illustrates what type of in vitro curve is required in order to develop an IVIVC for the Type 2 dual peak plasma profile. Although a significant amount of time has been spent in trying to develop an IVIVC and IVR system for formulations with a Type 2 profile, at the present time an example of this type of in vitro system or a validated IVIVC model describing both plasma peaks has not been reported in the literature.

For Type 2 plasma profiles, investigators have also attempted to develop an IVIVC for different parts of the plasma profile. This approach has been difficult because the first peak and second peak represent different release mechanisms

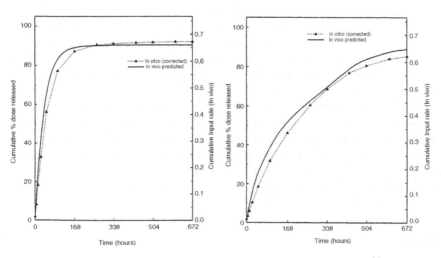

Figure 3 In vitro cumulative release and in vivo cumulative input of ^{14}C-methylated lysozyme from microspheres. *Source*: From Ref. 7.

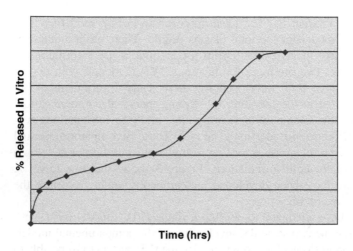

Figure 4 Plot percent released in vitro versus time that is required to develop an in vivo–in vitro correlation with Type 2, two peak plasma concentration profiles.

from the dosage form and the magnitude of the second peak appears to be related to the magnitude of the first peak. Although acceptable IVIVC models have been developed for formulation development work, IVIVC models that meet the strict predictability criteria of the FDA IVIVC guidance have not been reported for the Type 2 plasma profile.

The major problem appears to be in developing the relationship between the first peak and an in vitro release profile. The in vitro profile often represents more than the release of drug associated with only the surface of the microsphere, usually the major source of drug for the first plasma peak.

IVIVC models describing the second peak (i.e., after the initial burst) have been successfully developed. Using a Level C correlation for in vitro and in vivo data after the initial burst, Blanco-Prieto et al. (8) were able to develop a correlation for more than two formulations (Fig. 5). The validity of the model, however, was not investigated for internal or external predictability. Time invariant, predictable Level A IVIVC models have also been successfully developed for the second peak. Table 1 illustrates how a time variant model successfully related the in vitro release over seven days to the in vivo release of a second peak that occurred 20 to 50 days after administration. The IVIVC model met the strict predictability criteria of the FDA IVIVC guidance for both the C_{max} and AUC of the second peak (D. Young, personal communication, 2002).

Liposomes

Since liposomes have played a minor role as a parenteral MR delivery system over the years, there are no examples of a Level A IVIVC within the literature.

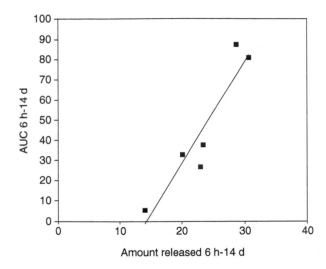

Figure 5 Relationship between the amount of peptide released after the initial burst until day 14 and the AUC of plasma calculated for the same period of time ($r = 0.932$). *Source*: From Ref. 8.

A good example showing the potential ability to develop a Level A IVIVC is illustrated in Figure 6 (9). Although the correlation between in vivo absorbed and in vitro release was for a single formulation, this example illustrates that the IVIVC methods reported within this book can be used for liposomes and that an IVIVC with more than two formulations may be possible for MR liposome drug delivery systems.

Implants

The development of IVIVC models for implants has received the most attention within the literature. Chilukuri and Shah (10) were able to develop a relationship between in vivo and in vitro release for a single vancomycin implant product

Table 1 Internal Validation of an In Vivo–In Vitro Correlation Model Relating the Second Peak to In Vitro Release for Three Microsphere Formulations (A,B,C) that Have Type 2 Plasma Profiles

Treatment	C_{max} %PE	AUC PE
A	11.8	6.7
B	9.5	13.4
C	2.1	5.3
MAPPE	7.8	8.5

Abbreviation: MAPPE, mean absolute percent prediction error.

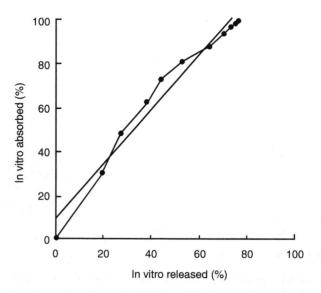

Figure 6 Correlation of percentage of drug absorbed in vivo and percent drug released in vitro for bre-MVL (triolein/tricaprylin, 10/0). *Source*: From Ref. 9.

(Fig. 7) and Baro et al. (11) were able to show a relationship for a single gentamicin formulation (Fig. 8).

One of the few literature reports demonstrating a relationship between in vitro and in vivo release for more than one formulation has been reported by Negrin et al. (12). The investigator manufactured implants and with slight differences in their in vitro release profiles (Fig. 9) evaluated the in vivo performance

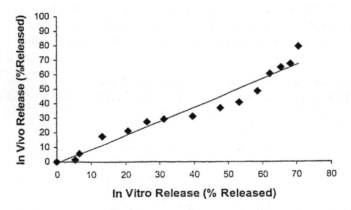

Figure 7 Correlation of in vitro–in vivo release of vancomycin from GMS implants ($R^2 = 0.97$, slope = 0.94). *Source*: From Ref. 10.

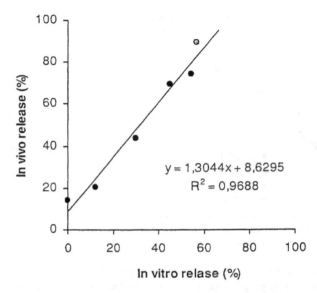

Figure 8 In vitro–in vivo gentamicin sulfate release correlation obtained for F-D. *Source*: From Ref. 11.

in animals. Negrin et al.'s investigation showed that all three implants followed a common in vitro to in vivo relationship (Fig. 10). Unfortunately, the authors did not provide any information about an IVIVC mathematical model nor the predict-ability of the in vitro–in vivo relationship.

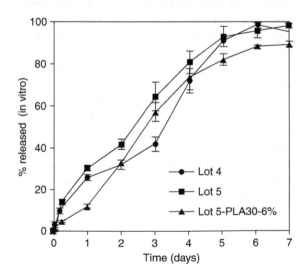

Figure 9 In vitro release profile from different methadone base implant designed for 1-week release. *Source*: From Ref. 12.

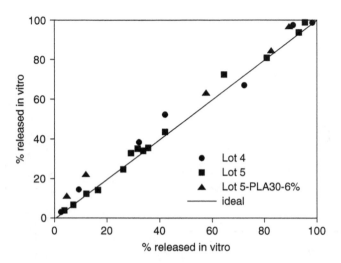

Figure 10 In vitro–in vivo correlation obtained for different 1-week methadone base implants. The in vivo release percentage represented here was calculated from the remaining methadone in the implant. *Source*: From Ref. 12.

CONCLUSION

The FDA IVIVC MR guidance for oral products has been used to develop IVIVC models for parenteral products. This chapter has provided some of the specific aspects of IVIVC modeling that are relevant to MR parenteral drug delivery systems as well as selected examples to illustrate the possibility of developing an IVIVC for a parenteral product. Given the number of different parenteral delivery systems, it is not possible to discuss or present examples for each system but the basic principles presented here apply to all parenteral delivery systems. Within the literature there are a number of examples demonstrating that a relationship exists between the in vitro release of a single formulation and the in vivo release. There are few examples, however, of the in vitro to in vivo relationship being demonstrated for more than one formulation. In addition, the development of mathematical models that describe the relationships have not been provided for most of the single or multiple formulation examples within the literature. Although the actual IVIVC mathematical models have not been reported in the literature, the examples within this chapter demonstrate that it is possible to develop an IVIVC model for parenteral products that can be used throughout the development and regulatory cycle.

ACKNOWLEDGMENT

The authors would like to thank Wendy Guy for her assistance in preparing the chapter.

REFERENCES

1. AAPS workshop. Assuring Quality and Performance of Sustained and Controlled Release Parenterals. Washington DC, April 19–20, 2001.
2. Food & Drug Administration Guidance for Industry. Extended release oral dosage forms: development, evaluation, and application of in vitro/in vivo correlations — CDER 09/1997.
3. Gillespie WR. Advances in experimental medicine and biology. In Vitro-In Vivo Correlations. Vol. 23, ch. 5. In: Convolution-Based Approaches for in Vivo–in Vitro Correlation Modeling. New York: Plenum Press, 1997.
4. Young D. In vitro–in vivo correlations in drug development workshop: clarifying the FDA perspective on IVIVC analysis based on interactions and submissions. Brussels, Belgium, 2003.
5. Bigora S, Farrell C, Shepard T, Young D. IVIVC applied workshop manual— Principles & Hands-on applications in pharmaceutical development. PDx by Globo-Max, 2002.
6. Devane J. Advances in experimental medicine and biology. In Vitro-In Vivo Correlations. ch. 23. In: Impact of IVIVR on Product Development. New York: Plenum Press, 1997.
7. Van Dijkhuzien-Radersma R, Wright SJ, Taylor LM, John BA, de Groot K, Bezemer JM, et al. In vitro-in vivo correlation for ^{14}C-methylated lysozyme release from poly (ether-ester) microspheres. Pharm Res 2004; 21(3):484–491.
8. Blanco-Prieto MJ, Campanero MA, Besseghir K, Heimgatner F, Gander B. Importance of single or blended polymer types for controlled in vitro release and plasma levels of a somatostatin analogue entrapped in PLA/PLGA microspheres. J Control Release 2004; 96:437–488.
9. Zhong H, Deng Y, Wang X, Yang B. Multivesicular liposome formulation for the sustained delivery of breviscapine. Int J Pharm 2005; 301:15–24.
10. Chilukuri D, Shah J. Local delivery of vancoymycin for the prophylaxis of prosthetic device-related infections. Pharm Res 2005; 22(4):563–572.
11. Baro M, Sánchez E, Delgado A, Perera A, Évora C. In vitro-in vivo characterization of gentamicin bone implants. J Control Release 2002; 83:353–365.
12. Negrín CM, Delgado A, Llabrés M, Évora C. Methadone implants for methadone maintenance treatment. In vitro and in vivo animal studies. J Control Release 2004; 95:413–421.

8

In Vitro–In Vivo Correlation: Transdermal Drug Delivery Systems

Ayyappa Chaturvedula
Glaxo SmithKline, Parsippany, New Jersey, U.S.A.

Ajay K. Banga
College of Pharmacy, Mercer University, Atlanta, Georgia, U.S.A.

INTRODUCTION

Transdermal delivery provides a potential noninvasive route of drug delivery, but only a small number of products are on the market. Skin, being a protective organ of the body, does not allow permeation of significant quantities of foreign substances into the body and the main barrier is the stratum corneum (SC). Only small and more lipophilic molecules could be delivered through skin until recently. However, the invention of novel transdermal delivery technologies such as iontophoresis, sonophoresis, electroporation, and microporation can now allow the delivery of hydrophilic small and macromolecules. Still, the commercial application of these technologies is difficult because of difficulties such as miniaturizing the devices. Iontophoresis has gained considerable importance in commercial development for systemic delivery of drugs due to the advantages of controlled and programmable delivery, and as a result some iontophoretic devices are in late stages of clinical development. An iontophoretic device has been approved by Food and Drug Administration for systemic delivery of fentanyl and another for local iontophoretic delivery of lidocaine. In vitro–in vivo correlation (IVIVC) in the case of transdermal drug delivery (TDD) is somewhat different compared to the oral route of administration. In most instances, steady state flux has been extrapolated to determine the in vivo plasma concentrations using pharmacokinetic (PK) modeling. IVIVC can help to achieve a mechanistic understanding of the drug delivery system. Unfortunately, until recently, the transdermal route has not been given enough attention as a potential delivery system by the industry even though a lot of academic research is being done. There are very few reports on IVIVC of transdermal delivery. The objective of this review is to make the reader aware of the theoretical aspects of skin permeation in passive and iontophoretic drug delivery systems, novel in vitro models for skin permeation studies, PK models that describe the transdermal delivery, and literature reports of IVIVC of TDD. This information will help further development of IVIVC of TDD systems.

STRUCTURE OF SKIN

Skin is the largest organ in the body and accounts for more than 10% of body mass. It consists of epidermis, dermis, and subcutaneous tissue. The SC is composed of the outermost nonviable cell layers of epidermis and is approximately 10 to 20 μm thick. It is composed of 15 to 25 flattened, stacked, hexagonal, and cornified cells embedded in intercellular lipid. Each cell in SC is approximately 40 μm in diameter and 0.5 μm in thickness. Dermis is about 0.1 to 0.5 cm thick and composed of collagenous fibers and elastic connective tissue matrix of mucopolysaccharides. Dermis contains an extensive vascular network with a blood flow rate of 0.05 mL/min/cc and is the layer responsible for systemic absorption of drugs delivered through the skin (1–4). Appendages such as hair follicles lie deep in the dermis.

TRANSDERMAL DRUG DELIVERY—PRINCIPLES OF DIFFUSION

Passive Transdermal Delivery

For drugs to be absorbed through the skin, they should be dissolved in the vehicle from which they will partition out, then permeate through the SC, and reach the systemic circulation via absorption by the blood capillaries under the epidermis. Permeation through the SC is usually the slowest process and, therefore, the rate-limiting step. However, there are some transdermal patches available with rate-controlling membranes in which diffusion through the membrane or matrix is slower than skin permeation and thus will be the rate-limiting step. Passage of the drug molecules through the skin can be described by Fick's laws of diffusion. The flux of a drug through the skin is dependent on the saturation solubility of the drug in the formulation vehicle. Flux can be described by the following equation:

$$J = \frac{DK}{h}(C_v - C_b) \tag{1}$$

where D is the diffusion coefficient of the drug through the skin, K the partition coefficient, h the thickness of the skin, C_v the solubility of drug in the vehicle, and C_b the plasma concentration.

Iontophoretic Drug Delivery

Iontophoresis involves the delivery of ionic drug molecules under electric fields. The drug molecules will take the least resistance pathways, such as through hair follicles. There are two components to the iontophoretic delivery: electro-repulsion and electro-osmosis.

$$J_{\text{iontophoresis}} = J_{\text{electrorepulsion}} \pm J_{\text{electro-osmosis}} \tag{2}$$

where J refers the flux of the drug molecule in each component, electrorepulsion is the simple repulsive force between similar charges, and electro-osmosis is the convective solvent flow in the direction of anode to cathode because of the current passage.

Skin is negatively charged at physiological pH and acts as a cation-selective membrane for the permeation. The direction of electro-osmosis can be changed by altering the membrane's electrical properties by preferential binding of cations to the fixed negative charges in the membrane (5). Electro-osmosis facilitates the passage of cationic species, but inhibits that of anionic species. It can be used to deliver polar and neutral solutes. The relative importance of electrorepulsion and electro-osmosis depends on the physicochemical and electrical characteristics of the membrane (6).

Theories of Electrotransport

Iontophoretic transport of monovalent cations and anions can be approximately predicted by the modified Nernst–Planck model, which is modified to account for the influence of convective solvent flow.

The steady state flux ($J_{\Delta\psi}$) during iontophoresis in a porous membrane can be written as follows:

$$J_{\Delta\psi} = \varepsilon\left\{-\left(D\frac{dC}{dX} + \frac{CzF}{R_{\text{gas}}T}\frac{d\psi}{dx}\right) \pm vC\right\} \tag{3}$$

where ψ is the electric potential in the membrane; F, Faraday's constant; R_{gas}, the gas constant; T, the temperature; v, the average velocity of the convective flow; ε, the combined porosity and tortuosity factor of the membrane; C, the concentration; x, the position in the membrane; and z, the charge number.

The predictions of the fluxes from this equation are consistent with the induction of pores in iontophoresis. A better appreciation of the meaning of the equation may be achieved by considering the case of a nonelectrolyte in which the charge $z = 0$, and Equation 3 is reduced to:

$$J_{\Delta\psi} = \varepsilon\left\{-\left(D\frac{dC}{dX}\right) \pm vC\right\} \tag{4}$$

This is similar to Fick's first law of diffusion with the porosity and tortuosity of the membrane and the effective convective flow added in the equation.

In the case where the permeant molecular radius is of the order of magnitude of the membrane pore radius, hindrance considerations must be included and Equation 3 can be rewritten as follows:

$$J_{\Delta\psi} = \varepsilon\left\{-\left(HD\frac{dC}{dX} + \frac{CzF}{R_{gas}T}\frac{d\psi}{dx}\right) \pm WvC\right\} \tag{5}$$

where H is the hindrance factor for simultaneous Brownian diffusion and migration driven by electric field and W is the hindrance factor for permeant transport via convective solvent flow. Equation 5 is called modified Nearnst–Plank equation (7).

The steady state flux for passive transport ($J_{passive}$) across a porous membrane can be expressed as:

$$J_{passive} = \frac{DC_D\varepsilon H}{\Delta x} \tag{6}$$

where H is the hindrance factor for passive diffusion; C_D, the concentration of the solute on the donor side; and Δx, the effective thickness of the membrane.

After integration of Equation 3, the steady state flux can be expressed as:
For anions

$$J_{\Delta\psi} = \frac{C_D\varepsilon HD\{[Wv/(HD)] + [K/\Delta x]\}}{\exp\{[Wv(\Delta x)/(HD)] + K\} - 1} \tag{7}$$

For cations

$$J_{\Delta\psi} = \frac{C_D\varepsilon HD\{[Wv/(HD)] - [K/\Delta x]\}}{1 - \exp\{[K - Wv(\Delta x)/(HD)]\}} \tag{8}$$

where $K = (zF\Delta\psi)/(R_{gas}T)$ and $\Delta\psi$ is the applied voltage across the membrane.

Enhancement factor (E_{total}) due to iontophoresis can be determined by the following equation:

$$E_{total} = \frac{J_{\Delta\psi}}{J_{passive}} \tag{9}$$

In another study by Li et al., the average effective pore sizes induced by the electrical field was measured around $12°A$, which were of the same order of magnitude as those of pre-existing pores determined from the conventional passive diffusion experiments (8,9). Li et al. showed the correlation between the electromobilities predicted and observed by the model. They suggested that the modified Nernst–Planck models predictions are satisfactory only when the electromobilities and the effective molecular size of the molecules are known (10).

Kontturi and Murtomaki proposed a model with two penetration routes for iontophoresis: one aqueous and one lipoidal. The mathematical form of the model is as follows:

$$J_p = J^w + J^o \qquad (10)$$
$$J_{if} = E \times J_p + J^o \qquad (11)$$

where J^w and J^o are the fluxes across the aqueous and organic pathways, respectively; J_p and J_{if} are the total passive and iontophoretic fluxes, respectively; and E represents the iontophoretic transport (11). Hirvonen et al. confirmed the predictability of the model and confirmed the idea that hydrophilic drugs experience the greatest benefit from iontophoretic delivery, whereas the flux enhancement of lipophilic drugs remains low (12).

Manabe et al. proposed the hydrodynamic pore theory for iontophoretic drug transport. They assumed parallel permeation pathways, pore and lipid pathways. Pore pathways are the main routes for hydrophilic drugs, as explained by the pore theory, and the net flux of a drug at steady state (J) can be described as the sum of each pathway flux:

$$J = J_L + J_P \qquad (12)$$

where J_L and J_P represent the flux through lipid and pore pathways, respectively (13).

IN VITRO MODELS OF SKIN PERMEATION STUDIES

Franz Diffusion Cell Set Up and Novel Modifications

For in vitro transdermal and topical absorption studies, Franz diffusion cells are widely used compared to the other methods. They consist of a donor and receptor compartment with the membrane of interest clamped between the two compartments. They can be either static or flowthrough, based on how the receptor phase is replenished, and provide a cost-effective evaluation method. But the inherent problem with static Franz diffusion cells is the lack of microvasculature, which is present in the in-vivo situation and helps in rapid clearance of the drug. It is particularly a problem when poorly soluble drugs are tested by these methods. For poorly soluble drugs, the drug concentration may reach closer to saturation solubility and, thus, the assumption of sink conditions may no longer be valid. To avoid these problems, flowthrough cells were designed in such a way that the receptor buffer is continuously removed (14). Flux is calculated as the linear slope of the cumulative amount delivered versus the time profile in standard Franz diffusional cell experiments. Permeability coefficient (P), using the steady state flux (J), can be derived by the following equation:

$$J = A \cdot P \cdot \Delta C \qquad (13)$$

where A is the surface area and ΔC the difference in the donor and receptor concentration.

However, in the case of flowthrough diffusion cell experiments, the collector volume, receiver cell volume, flow rate, and sampling interval may modify the apparent flux data. Additionally, these parameters may influence the finite dose flux profiles and the infinite dose diffusional lag times. In general, the concentration in the receiver (C_r) of a flow-through diffusion cell versus time (t) in the case of infinite dose can be studied by the following equation (15):

$$V_{rec} \cdot \frac{dC_{rec}}{dt} = J \cdot A - F_{rec} \cdot C_{rec} \tag{14}$$

where V_{rec} is the volume of receiver chamber; J, the flux of drug out of the skin; A, the diffusional area; and F_{rec}, the flow rate of the receptor fluid.

In the case of finite dose studies, the flux increases to steady state flux and then decreases. Thus, the apparent flux profiles can be fitted to a bi-exponential function by the following equation:

$$J = A \cdot \left(e^{-K_1 \cdot t} - e^{-K_2 \cdot t}\right) \tag{15}$$

where K_1 and K_2 represent the apparent increasing and decreasing rate constants of J. The first derivative of Equation 15 gives the value of maximum J, corresponding to the flux at steady state (16).

Tanojo et al. proposed a modification to the flowthrough diffusional cells. In the modified cells, the contact between the membrane and the receptor chamber was optimized by a spiral channel, which formed when the membrane was clamped between the donor and receptor compartments. Thus, the concentration of donor can be ensured throughout the experimental period by a separate flowthrough system. These cells have the unique advantage of using both finite and infinite doses in the donor. This design eliminates any stagnant domains in the compartment and sink conditions, and proper mixing in the receptor compartment is assured. The cumulative amounts reached plateau for p-amino benzoic acid through human SC in the case of a finite dose, due to the depletion of the drug in three hours, whereas it increased throughout the 16 hours that the constant donor concentrations were maintained (17). Bosman et al. developed Kedler cells, which can be used in combination with the automatic sample preparation with extraction columns system. These are the automated alternative to the static Franz cells. These cells consist of inlet, donor, and receptor compartments. This design has advantages of automated sampling, reduction of air bubble entrapment, mimicking the in vivo situation by the continuous flow of receptor buffer, and smaller areas of skin can be used for experiments. They showed the comparative results of Kedler cells with Franz cells for the permeation of dexetimide (18,19).

Isolated Perfused Porcine Skin Flap

Riviere et al. developed an isolated perfused porcine skin flap (IPPSF) model, which consists of porcine skin isolated with the microcirculation. This method can be used to determine in vitro absorption and metabolism within the skin. Pigskin has the potential disadvantage of excessive fat, and this allows the drug to distribute into the fat rather than into the actual sampling compartment (20). The possible necrosis of the excised skin during the long experimental conditions may affect the in vitro flux of drugs. To avoid this problem, Venter et al. developed an in situ adapted diffusional cell model in nude mice, which they compared with the static Franz cell data. In this method, a modified diffusion cell was implanted under the dissected dorsal skin of the animal. The skin was dissected with the least bleeding possible and which represents minimal disturbance of blood supply to the specific skin area. The skin was clamped between the donor and receptor chambers and consisted of intact cutaneous microvasculature. Histological observations showed that necrosis occurred in both epidermal and dermal regions in the excised skin pieces. They observed a higher flux in vitro compared to the in situ situation and could be due to the structural changes in the excised skin (21).

IN VIVO MODELS OF SKIN PERMEATION STUDIES

A variety of animal models were used both for in vitro and in vivo skin absorption studies. Commonly used excised skin membranes for diffusion studies are from hairless rats, hairless mice, fuzzy rats, guinea pigs, rabbits, and human. Of all of the aforementioned cases, human skin is ideal, but because of availability and ethical constraints there is a need for a good animal model that can mimic human skin. All the animal species differ in hair follicle density per unit area and SC thickness. Generally, thickness of the SC increases with the size of the animal. There are several conflicting reports comparing different animal species for in vitro permeation, but the results show that the permeation is more subjective to the drug and no generalized conclusions can be made. Wester and Noonan. (22) suggested that pig and monkey were the most predictive models for percutaneous absorption in human. Niazy (23) observed that a species difference in penetration-enhancing effect of a zone and the enhancement factor was in the order rabbit skin > human skin > rat skin > guinea pig skin > hairless mouse skin. Panchagnula et al. investigated the permeation characteristics of a lipophilic and hydrophilic molecule across various animal models in vitro. By comparing the lag times and permeability values of both compounds, they suggested that the guinea pig, hairless guinea pig, and Brattleboro rat were good models for transdermal absorption studies. They did not find any qualitative or quantitative relationship between number of hair follicles and permeability of either of the compounds (24). Artificial composite membranes composed of silicone and poly 2-hydroxy ethyl methacrylate were also investigated as an alternative to skin membranes. Composite membrane permeabilities for a wide range of drugs with diverse physicochemical properties were compared and reasonable correlation was found between the calculated and observed permeability coefficients (25). Shed snakeskin was considered to have a close correlation with the permeation rates of drugs through human SC and they are devoid of hair follicles, and thus can be used for mechanistic studies. But the thickness and permeability characteristics differ in different species of snakes and thus they showed that the snakeskin is not a good model for human skin (26). Rittirod et al. demonstrated metabolic differences in the in vitro skin experiments. They showed that the differences in permeation profiles of ethyl nicotinate among the species tested was primarily due to the differences in the esterase activity, and thus when extrapolating the skin permeation data to human skin, skin metabolism should be considered (27). Hikima et al. (28) demonstrated the differences in hydrolase activity in different animal species in vitro and they suggested that the skin of the hairless rat and the hairless mouse might be good models to simulate clinical situation of hydrolytic activity. Recently, cultured skin was used as alternative to the excised animal skin. Excised animal skin has the disadvantages of high variability and a lack of viability. Suhonen et al. investigated an epidermal cell culture derived from rat keratinocytes for permeability characteristics. They showed that the organitryptic ketinocyte culture may give a very close estimate on human epidermal membrane permeabilities over a large range of lipopholicities and molecular weights (29). These cultures are devoid of appendages and, thus, may not be useful to investigate iontophoretic drug delivery. Monteiro-Riviere et al. used the human epidermal keratinocytes to demonstrate an irritant induced cytokine release. This method has a good potential to investigate the drug-skin interactions (30). Grafe et al. demonstrated a pH dependent carrier mediated transport of clonidine in human keratinocytes cultures. They proposed that after the passage through the SC, the transporter might contribute to the passage of clonidine (31).

PHARMACOKINETIC MODELS FOR TRANSDERMAL DRUG DELIVERY

Drug delivery through skin is a complex phenomenon due to the complexity involved in the skin structure. The ultimate aim of the PK modeling is to predict the future state of a system, which plays a major role in optimizing the dosing regimens and toxicological consequences.

Diffusion Model

Berner derived a PK model for passive diffusion of drug from the device with finite and infinite doses. In Figure 1, the transdermal device consists of a stirred-drug reservoir and a rate controlling membrane in contact with the SC. It was assumed that the membrane is in equilibrium with the drug reservoir. Collection of urine sample after transdermal patch application is the easiest and noninvasive way to monitor the fate of the drug in the body.

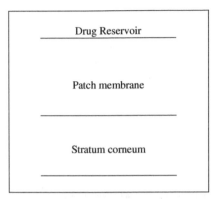

Figure 1 Transdermal devices with infinite dose. *Source*: Adapted from Ref. 32.

Total amount excreted in urine (M_u) can be related to the steady state flux (J_{ss}) as

$$M_u^a(t) = \frac{k_u A J_{ss}}{K}(t - t_L^D - 1/K) \qquad (16)$$

where M_u^a is the asymptotic limit of M_u; k_u, the urinary excretion rate; A, the surface area of the patch; K, the elimination rate constant; t, the time; and t_L^D, the lag time to reach steady state diffusion.

Steady state blood concentrations can be obtained as

$$C_b = \frac{A J_{ss}}{K V_d} \qquad (17)$$

where C_b is the blood concentration and V_d the volume of distribution.

PK lag time (t_L) can be calculated as

$$t_L = \frac{1}{K} + \frac{l_s^2}{6 D_s} \qquad (18)$$

where l_s is the thickness of SC, and D_s the diffusion coefficient of the drug through SC

PK lag time is composed of PK contribution, which is the lag time for diffusion across the SC. This is precisely the result for the PK lag time for the skin permeation alone (32).

Parallel Infusion Model

This model assumes that transdermal patch provides two simultaneous drug infusions resulting from the burst effect and parallel controlled maintenance infusion. In Figure 2, the transdermal patch has a temporary compartment due to the burst

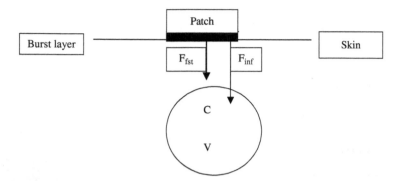

Figure 2 Pharmacokinetic model with parallel infusions. *Abbreviations*: F_{fst}, fast infusion from the burst layer; F_{inf}, slower maintenance infusion; C, concentration of distribution of plasma compartment; V, volume of distribution of plasma compartment. *Source*: Adapted from Ref. 33.

effect other than drug reservoir. The burst infusion (F_{fst}) and the maintenance infusion (F_{inf}) will contribute to the total drug input into the body.

Plasma concentration (C) can be represented as:

$$V \cdot \frac{dC}{dt} = F_{inf} + F_{fst} - Cl \cdot C \tag{19}$$

Cl and V represent total body clearance and volume of distribution, respectively.

F_{fst} and F_{inf} are defined as:

$$F_{fst} = \frac{Dose_{fst}}{Time_{fst}} = \frac{Released\ dose - Dose_{inf}}{Time_{fst}} \tag{20}$$

$$F_{inf} = \frac{Dose_{inf}}{Time_{inf}} \tag{21}$$

where $Time_{fst}$ and $Time_{inf}$ are the times of burst and maintenance infusions; $Dose_{inf}$ and $Dose_{fst}$ are the doses delivered in maintenance and burst infusion periods (33).

Multicompartment Model Describing Intradermal Kinetics

Nakayama et al. developed a six compartmental model considering skin and muscle layers as different compartments. They showed that the skin and muscle layer below the patch act as a different compartment compared to the contralateral skin and muscle layers. In Figure 3, V and C represent volume of distribution and drug concentration in each compartment, respectively. Cl indicates the clearance between the tissues and tissue or plasma. The subscripts d, vs, m, and p reveal the donor cell, viable skin, muscle, and plasma compartments, respectively.

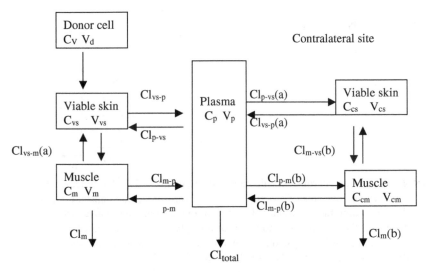

Figure 3 Six compartmental model for intradermal disposition kinetics. *Abbreviations*: C, concentration; V, volume; Cl, clearance; Subscripts vs, m, p, cm, and cs, viable skin, muscle, plasma, contralateral muscle and contralateral skin; a $= V_{cs}/V_s$ and b $= V_{cm}/V_m$. *Source*: Adapted from Ref. 34.

Differential equations for the compartments described earlier give the drug distribution in each tissue compartments. In this model, the values of Cl_{vs-m} and Cl_{vs-p} could be the indices for the efficiency of the systemic delivery or the local delivery and they showed that 20% antipyrine in the viable skin after topical application can directly penetrate into the muscle layer below the application site. This model could be used to evaluate the fractional contribution of the direct penetration and the blood supply to the deeper muscle layer below the patch application site (34).

Population Pharmacokinetic Model

Auclair et al. proposed a PK model to calculate population PK parameters of nitroglycerin and its metabolites after transdermal administration. They used a one-compartment model with a first-pass, mixed-order release (Fig. 4). This model showed a very good fit to the observed data following transdermal nitroglycerin in human volunteers. Due to the reservoir and membrane in the transdermal device, a mixed-order release gave a better fit than the assuming first-order or zero-order release alone. A lag time is needed in this model because time is needed for equilibrium to be made between the patch and the skin, and for the skin reservoir to be filled before the appearance of the drug in the systemic circulation. This model has an advantage in a head-to-head comparison between different formulations of nitroglycerin by minimizing the errors

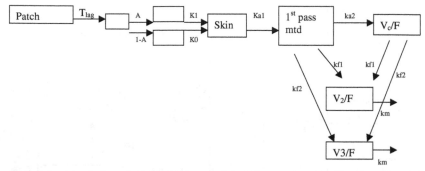

Figure 4 Population pharmacokinetic model. *Abbreviations*: T_{lag}, lag time; K_1, K_{a1}, and K_{a2}, first-order rate constants of absorption; K_0, the zero-order rate constant; k_{f1} and k_{f2}, metabolite formation rate constants; k_m, metabolite elimination rate constant; V_c/F, V_2/F, and V_3/F, the volume of distribution of nitroglycerine and its metabolites; A, the percentage of drug reaching systemic circulation that was released from the patch by a first order process (K_1). *Source*: Adapted from Ref. 35.

resulting from the fluctuations in the plasma concentrations and differentiating the formulations in terms of onset of action and drug elimination (35).

Physiologically Based Pharmacokinetic Model

Tashiro et al. developed a modified physiologically based PK (PBPK) model to show the effect of iontophoresis on drug transfer from skin to cutaneous blood. It has been shown that cathodal iontophoresis of ketoprofen resulted in a higher cutaneous plasma concentration than the systemic concentration, demonstrating the transfer of ketoprofen from skin to cutaneous blood. Transfer rate of the drug from skin to cutaneous blood (R_{sc}) can be defined as:

$$R_{sc} = Q_b(C_{vein} - C_{artery}) \tag{22}$$

where Q_b is the flow rate of cutaneous blood; C_{artery}, the drug concentration in the arterial blood entering cutaneous circulation; and C_{vein}, the drug concentration in the venous blood leaving cutaneous circulation.

In Equation 22, C_{artery} and C_{vein} can be substituted by blood and plasma concentration in systemic vein and cutaneous vein ($C_{b,systemic}$, $C_{p,systemic}$ and $C_{b,cutaneous}$, $C_{p,cutaneous}$, respectively) and Equation 22 can be rewritten as:

$$R_{sc} = Q_b \frac{C_b}{C_p}\left(C_{p,cutaneous} - C_{p,systemic}\right) \tag{23}$$

This model can be used for kinetic analysis of local disposition in TDD. They showed that iontophoresis induced the change of drug transfer to cutaneous blood flow and the change may depend on the application period and magnitude of electrical current (36).

Convolution/Deconvolution Methods

Deconvolution can be used to determine the input function into the systemic circulation. Bioavailability can be estimated by the integral of this input function. If the disposition of the drug after intravenous bolus administration is $w(t)$ and input rate is $I(t)$, the response function $r(t)$ can be considered as a cumulation of various intravenous doses, assuming that the presence of drug does not affect its kinetics. Thus, the relation between the input rate, intravenous bolus or unit impulse response, and response function can be written as the convolution function (Eq. 24):

$$C = r(t) = I(t) * w(t) \tag{24}$$

where C is the plasma concentration and equal to $r(t)$ (the response to the input function) and $*$ the convolution operator (33).

Deconvolution is the opposite mathematical operation of convolution and can be described by Equation 25.

$$I(t) = r(t)//w(t) \tag{25}$$

where $//$ is the deconvolution operator.

Phase Plane Methodology

Dokoumetzidis and Macheras investigated the phase plane method to characterize absorption kinetics of drug delivery systems. In this method, phase plane plots consist of dC/dt versus concentration (C) using concentration (C) and time (t) data. The shape of the phase plane plot is indicative of output and input kinetics. Input rate can be related to concentrations:

$$\text{Input} = \frac{dC}{dt} + K_e C \tag{26}$$

where K_e is the first-order elimination rate constant.

They calculated the ratio of the slopes of absorption and elimination phases and compared to the value 2, a theoretical minimum for the ratio of the slopes. Zero-order absorption kinetics will have a patent discontinuity between absorption and elimination phases, and similar slopes for the data above and below the x-axis. By modeling errant data, they showed that the ratio of slopes (r) indicates the absorption kinetics. A $r < 2$ represents the presence of zero order kinetics and $r > 2$ represents the presence of first order of Michaelis–Menten kinetics. Additionally, this method allows for the construction of fictitious "cumulative concentration-time" profiles by calculating the area under the curve of the input rate for each time interval. When each of these quantities is divided by the total area, this provides the "normalized fictitious concentration-time" profile, which has all the qualitative features of the classical percent absorbed versus time plots (37).

IN VITRO–IN VIVO CORRELATION OF TRANSDERMAL DRUG DELIVERY

Some focused published examples to predict in vivo blood levels using the models described earlier will provide more practical insight into the utility of the models. Examples for passive and iontophoretic drug delivery will be discussed as separate categories.

Passive Transdermal Delivery

Physiologically Based Pharmacokinetic Model

Macpherson et al. developed a PBPK model (Fig. 5) consisting of four compartments: (*i*) rapidly perfused tissues, (*ii*) slowly perfused tissues, (*iii*) liver (major route of elimination), and (*iv*) plasma to predict the in vivo blood levels of benzoic acid (BA) after transdermal delivery. Schematic representation of the model is given subsequently. In vitro percutaneous absorption studies were performed using flowthrough diffusion cells and the skin was perfused with oxygenated HEPES-buffered Hank's balanced salt solution, pH 7.4 at a

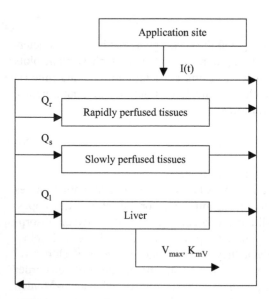

Figure 5 Physiologically based pharmacokinetic model developed to predict plasma levels of benzoic acid (BA) in hairless guinea pigs after topical exposure to a finite dose. Topically applied BA is absorbed into the systemic circulation; the rate of percutaneous absorption is represented by the transdermal input function I(t). Absorbed BA is distributed by blood or plasma to the compartments represented as rapidly perfused tissue, slowly perfused tissue, and liver. *Source*: From Ref. 38.

flow rate of 1.6 mL/hr to keep the tissue viable. The transdermal input rate $I(t)$ was calculated from the following first-order, one-compartment absorption model:

$$I(t) = \frac{\left[K * K_a * \alpha \left(e^{-K*t} - e^{-K_a*t}\right)\right]}{K_a - K} \tag{27}$$

where K is the elimination rate constant from skin; K_a, the absorption rate into the skin; and α, the scaling parameter. The parameters in the model were calculated by fitting the in vitro data to the above model using SAS computer program. Differential equations for the PBPK model in different tissues were written and the blood levels in the plasma compartment can be written as:

$$V_b * \left(\frac{dC_b}{dt}\right) = \left[\frac{(Q_r * C_r)}{P_{(r/b)}}\right] + \left[\frac{(Q_s * C_s)}{P_{(s/b)}}\right] + \left[\frac{(Q_1 * C_1)}{P_{(1/b)}}\right]$$
$$+ I(t) - (Q_b * C_b) \tag{28}$$

where C_b, C_r, C_s, and C_1, and Q_b, Q_r, Q_s, and Q_1 represent the concentration of BA in blood, rapidly perfused tissues, slowly perfused tissues, liver, and blood flow to the respective compartments. Partition coefficient between the different tissue compartments were described by $P_{r/b}$, $P_{s/b}$, and $P_{1/b}$. Volume of tissue compartments were either calculated experimentally or obtained from the literature. Partition coefficients were calculated by experimental methodology. The input rate $I(t)$ calculated from the in vitro experiment was utilized in Equation 28. Observed and predicted plasma levels of BA after exposure to three different dose levels are provided in Figure 6. The developed model was capable of predicting plasma BA levels that would result from topical exposure to finite doses of BA (38).

Two-Compartment Model

A two-compartment model with zero-order absorption was used to predict in vivo concentrations of a synthetic cannabinoid after transdermal delivery in guinea pigs. The patches were applied until the steady state plasma concentrations were achieved. Basic PK parameters were calculated from the plasma concentration-profile after intravenous administration of the cannabinoid. Steady-state plasma concentrations were calculated from the following equation:

$$C = \frac{J_{ss}S}{V_c K_{el}} \left[1 + \frac{\beta - K_{el}}{\alpha - \beta} \exp(-\alpha t) + \frac{K_{el} - \alpha}{\alpha - \beta} \exp(-\beta t)\right] \tag{29}$$

where α and β are the distribution constants; V_c, the volume of distribution of the central compartment; and K_{el}, the elimination rate constant obtained from intravenous administration. J_{ss} is the steady state flux observed in the in vitro permeation studies using hairless guinea pig skin. The predicted concentrations by the equation are shown in Figure 7 and a good correlation was observed (39).

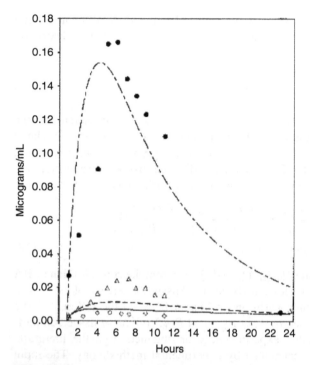

Figure 6 Predicted plasma BA levels with K_m and V_{max} optimized simultaneously at all dose levels and experimental data for topical exposures of 12, 40, and 120 µg of BA/cm^2. *Note*: Actual points represent observations and the *dashed lines* represent predictions by the model. *Source*: From Ref. 38.

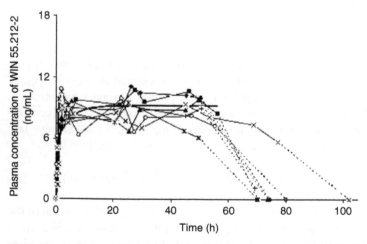

Figure 7 Individual plasma concentration profiles after transdermal application of cannabinoid. *Note*: *Dashed lines* represent the concentrations after patch removal and *solid line* represents the model predicted concentrations. *Source*: From Ref. 39.

Convolution Method

Qi et al. predicted in vivo plasma concentrations of 2,3,5,6-tetramethylpyrazine (TMP), following transdermal application by the convolution methodology. The unit impulse parameters A, B, a, b were calculated by intravenous administration of the compound and model fitting. A weighting function was estimated from the in vitro permeation data. TMP release showed a biphasic pattern with a burst release phase during the first four hours followed by a slow release phase from four to six hours, and further slower release phase from 6 to 24 hours. In order to include the effect of diffusion changes, a multiple-constant input in convolution procedure was followed. Convolution was performed by the following equations for all the phases:

If $0 \le t \le t$

$$C(t) = \frac{J_1 A}{a}(1 - e^{-at}) + \frac{J_1 B}{b}(1 - e^{-bt}) \tag{30}$$

If $t_1 \le t \le t_2$

$$C(t) = \frac{J_2 A}{a}\left[1 - e^{-a(t-t_1)}\right] + \frac{J_2 B}{b}\left[1 - e^{-b(t-t_1)}\right]$$
$$+ \frac{C(t_1)}{A+B}\left[Ae^{-a(t-t_1)} + Be^{-b(t-t_1)}\right] \tag{31}$$

If $t_2 \le t \le t_3$

$$C(t) = \frac{J_3 A}{a}\left[1 - e^{-a(t-t_2)}\right] + \frac{J_3 B}{b}\left[1 - e^{-b(t-t_2)}\right]$$
$$+ \frac{C(t_2)}{A+B}\left[Ae^{-a(t-t_2)} + Be^{-b(t-t_2)}\right] \tag{32}$$

The predictability of the model was evaluated using percent prediction error, and Figure 8 shows the correlation between predicted and observed concentrations of TMP in rabbit (40).

Iontophoretic Drug Delivery

IVIVC of iontophoretic delivery is more controversial due to the conflicting reports. Generally, a higher in vivo flux has been reported than in vitro flux in various reports. Most of these differences were ascribed to the intrinsic ion concentration of the skin. This could be due to the faster in vivo clearance of drug from the skin due to the cutaneous microcirculation and the presence of an ion reservoir in the skin that may compete for charge transport with the drug and the depletion of this reservoir in turn may contribute to the lag-time before a steady iontophoretic flux is achieved. Conventionally used Franz-diffusion cells have relatively smaller area than the actual patch studies in vivo. The amount of drug adsorbed per unit area could have a negative impact on the drug flux (41). Geest et al. studied the

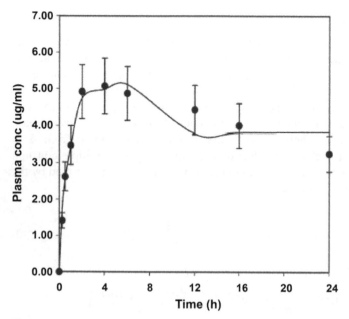

Figure 8 Transdermal delivery of 2,3,5,6-tetramethylpyrazine (TMP) after a single 24-hour application of TMP-TTS (300 mg/30 cm^2). *Note: Solid data points* represent the observations and the *line* represents the concentrations predicted by the convolution method. *Source*: From Ref. 40.

iontophoretic delivery of arbutamine and found that the IVIVC is highly dependent on the experimental conditions. Several mechanisms that contribute to the overall transport may effect in vitro and in vivo transport rates to a different degree (42).

Isolated Perfused Porcine Skin Flap Model to Predict In Vivo Concentrations

Riviere et al. utilized IPPSF model to predict iontophoretic delivery of arbutamine in humans. The in vivo PK parameters were calculated using an intravenous infusion of arbutamine in human. The observed concentrations of drug in the venous efflux from IPPSF versus time were normalized by the concentration of applied drug concentration, electronic dose, and electrode area. These input profiles were then renormalized by the in vivo conditions by multiplying with the concentration of drug in vivo, electron dose, and electrode area. The resultant input function was fed in the systemic model and thus these input profiles served as a continuous function and produced a simulated plasma concentration-time window with the following equation that was computed iteratively (two minute intervals):

$$M_t = \frac{U(t)}{K_{el}}(1 - e^{-K_{el}t}) + M_0(t)e^{-K_{el}t} \qquad (33)$$

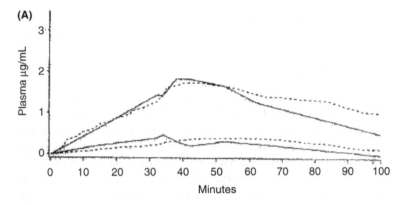

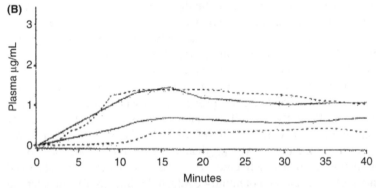

Figure 9 Observed and predicted drug concentration-time profiles resulting from ionto-phoretic percutaneous administration of drug to human subjects: (**A**) 32 minutes or (**B**) 10 minutes. *Note*: *Solid line* represents the observed concentrations and *dashed line* represents predicted concentrations. *Source*: Adapted from Ref. 43.

where M_t is the mass of drug in the body at time t; $U(t)$, the drug input (IPPSF output) rate; and $M_0(t)$, the initial drug mass determined from the previous iteration. Dividing $M(t)$ by steady-state volume of distribution yields the predicted concentration of drug in plasma. This model predicted the in vivo plasma concentrations well (Fig. 9). But the authors warn that the model must be validated for other drugs as the effect of drugs on the cutaneous blood circulation will in turn effect the over all results (43).

Cutaneous Microdialysis

The microdialysis technique to measure skin concentrations of drugs has been widely reported in the literature (44–48). It was portrayed as a minimally invasive method and more suitable than the current methods of drug assessment in the skin (49). Ault et al. investigated the effect of probe implantation in skin in vitro–in vivo histological examination and microdialysis delivery in fuzzy rats.

There was no inflammation or edema formation immediately after the probe insertion. However, increased infiltration of lymphocytes was observed after six hours of implantation and during the entire experiment (72 hours). Noticeable changes in the cells surrounding the probe were observed 32 hours of probe implantation (50). Groth and Serup investigated the probe insertion trauma on skin perfusion and erythema skin thickness in human. They reported that insertion with a 21 G needle increased blood flow and skin erythema, and baseline blood flow was recovered after 90 to 120 minutes. Ultrasound imaging studies showed a 38% relative skin thickening due to the probe (51). Similar experiments in hairless rats resulted in the probe placement around 1 mm in the skin and significant skin thickening (30%) was observed. The edema developed after 15 to 30 minutes after insertion and it was preferred to start microdialysis procedure after 30 minutes of probe insertion (52). Mathy et al. investigated the in vivo tolerance of skin after subcutaneous and dermal microdialysis probes in rat using bioengineering methods such as transepidermal water loss (TEWL), Laser Doppler velocimeter, and Chromameter. Dermal insertion of a probe using 26 G needle did not show any physical damage to the skin. Subcutaneous probe insertion did not show any significant changes in the skin characteristics. Elongation of the cells surrounding the probe was observed after 24 hours of probe insertion (53). Recently, we utilized subcutaneous microdialysis to show IVIVC of granisetron after iontophoretic delivery (54). We compared the area under the curve from time 0 to t (AUC_{0-t}) values of microdialysate data to the in vitro cumulative amounts delivered. A good correlation was observed (Fig. 10). This method could be potentially evaluated clinically for transdermal and topical bioequivalence of commercial drug products.

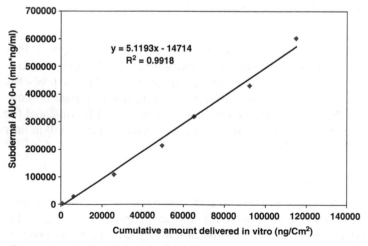

Figure 10 In vitro–in vivo correlation of iontophoretic delivery of granisetron using cutaneous microdialysis. *Source*: From Ref. 54.

CONCLUSION

IVIVC of transdermal delivery is more dependent on the experimental conditions and the nature of the molecule of interest, in case of Iontophoretic delivery, but is better established in passive transdermal delivery. However, the variability of human skin thickness and resulting differences in the permeation of drugs is a major problem that still needs to be addressed. It is sometimes hard to optimize the formulations due to inherent variability in the in vitro techniques. More comprehensive research should be done in this area to come up with better techniques to accurately predict the permeation characteristics of drugs from different formulation. Current in vitro methods are not sufficient for product development purposes but are good for candidate selection purposes. One major aspect of transdermal delivery is the irritation caused by the drugs. There is no reported method available, currently, for the in vitro method to evaluate the skin irritation potential of drug molecules, and researchers should be looking into this issue so that many more drug candidates could be screened for their irritancy potential in the drug discovery stages. More drug candidates could be marketable if TDD is also considered as a major route of drug delivery, as many drugs fail clinical development due to high first-pass metabolism.

ACKNOWLEDGMENT

We are grateful to Dr. C. Kolli for critically reading through the chapter and making helpful suggestions.

REFERENCES

1. Knepp VM, Hadgraft J, Guy RH. Transdermal drug delivery: problems and possibilities. Crit Rev Ther Drug Carrier Syst 1987; 4:13–37.
2. Moore L, Chien YW. Transdermal drug delivery: a review of pharmaceutics, pharmacokinetics, and pharmacodynamics. Crit Rev Ther Drug Carrier Syst 1988; 4:285–349.
3. Banga AK. Electrically Assisted Transdermal and Topical Drug Delivery. London: Taylor and Francis,1998.
4. Hadgraft J. Passive enhancement strategies in topical and transdermal drug delivery. Int J Pharm 1999; 184:1–6.
5. Burnette RR, Ongpipattanakul B. Characterization of the permselective properties of excised human skin during iontophoresis. J Pharm Sci 1987; 76:765–773.
6. Guy RH, Kalia YN, Delgado-Charro MB, Merino V, Lopez A, Marro D. Iontophoresis: electrorepulsion and electroosmosis. J Control Release 2000; 64:129–132.
7. Li SK, Ghanem AH, Peck KD, Higuchi WI. Iontophoretic transport across a synthetic membrane and human epidermal membrane: a study of the effects of permeant charge. J Pharm Sci 1997; 86:680–689.

8. Li SK, Ghanem AH, Peck KD, Higuchi WI. Characterization of the transport pathways induced during low to moderate voltage iontophoresis in human epidermal membrane. J Pharm Sci 1998; 87(1):40–48.
9. Higuchi WI, Li SK, Ghanem AH, Zhu H, Song Y. Mechanistic aspects of iontophoresis in human epidermal membrane. J Control Release 1999; 62:13–23.
10. Li SK, Ghanem AH, Teng CL, Hardee GE, Higuchi WI. Iontophoretic transport of oligonucleotides across human epidermal membrane: a study of the Nernst-Planck model. J Pharm Sci 2001; 90:915–931.
11. Kontturi K, Murtomäki L. Mechanistic model for transdermal transport including iontophoresis. J Control Release 1996; 41:177–185.
12. Hirvonen J, Murtomäki L, Kontturi K. Experimental verification of the mechanistic model for transdermal transport including iontophoresis. J Control Release 1998; 56:169–174.
13. Manabe E, Numajiri S, Sugibayashi K, Morimoto Y. Analysis of skin permeation-enhancing mechanism of iontophoresis using hydrodynamic pore theory. J Control Release 2000; 66:149–158.
14. Bronaugh RL, Stewart RF. Methods for in vitro percutaneous absorption studies IV: the flow-through diffusion cell. J Pharm Sci 1985; 74:64–67.
15. Cordoba-Diaz M, Nova M, Elorza B, Cordoba-Diaz D, Chantres JR, Cordoba-Borrego M. Validation protocol of an automated in-line flow-through diffusion equipment for in vitro permeation studies. J Control Release 2000; 69:357–367.
16. Kubota K, Yamada T. Finite dose percutaneous drug absorption: theory and its application to in vitro timolol permeation. J Pharm Sci 1990; 79:1015–1019.
17. Tanojo H, Bouwstra JA, Junginger HE, Bodde HE. In vitro human skin barrier modulation by fatty acids: skin permeation and thermal analysis studies. Pharm Res 1997; 14(1):42–49.
18. Bosman IJ, Lawant AL, Avegaart SR, Ensing K, de Zeeuw RA. Novel diffusion cell for in vitro transdermal permeation, compatible with automated dynamic sampling. J Pharm Biomed Anal 1996; 14:1015–1023.
19. Bosman IJ, Avegaart SR, Lawant AL, Ensing K, de Zeeuw RA. Evaluation of a novel diffusion cell for in vitro transdermal permeation: effects of injection height, volume and temperature. J Pharm Biomed Anal 1998; 17:493–499.
20. Riviere JE, Monteiro-Riviere NA. The isolated perfused porcine skin flap as an in vitro model for percutaneous absorption and cutaneous toxicology. Crit Rev Toxicol 1991; 21:329–344.
21. Venter JP, Muller DG, Du PJ, Goosen C. A comparative study of an in situ adapted diffusion cell and an in vitro Franz diffusion cell method for transdermal absorption of doxylamine. Eur J Pharm Sci 2001; 13:169–177.
22. Wester RC, Noonan PK. Relevance of animal models for percutaneous absorption. Int J Pharm 1980; 7:99–110.
23. Niazy EM. Differences in penetration-enhancing effect of Azone through excised rabbit, rat, hairless mouse, guinea pig and human skins. Int J Pharm 1996; 130(2):225–230.
24. Panchagnula R, Stemmer K, Ritschel WA. Animal models for transdermal drug delivery. Methods Find Exp Clin Pharmacol 1997; 19:335–341.
25. Hatanaka T, Inuma M, Sugibayashi K, Morimoto Y. Prediction of skin permeability of drugs. I. Comparison with artificial membranes. Int J Pharm 1990; 38(12):3452–3459.

26. Haigh JM, Beyssac E, Chanet L, Aiache JM. In vitro permeation of progesterone from a gel through the shed skin of three different snake species. Int J Pharm 1998; 170:151–156.

27. Rittirod T, Hatanaka T, Uraki A, Hino K, Katayama K, Koizumi T. Species difference in simultaneous transport and metabolism of ethyl nocotinate in skin. Int J Pharm 1999; 178:161–169.

28. Hikima T, Yamada K, Kimura T, Maibach HI, Tojo K. Comparison of skin distribution of hydrolytic activity for bioconversion of beta-estradiol 17-acetate between man and several animals in vitro. Eur J Pharm Biopharm 2002; 54:155–160.

29. Marjukka ST, Pasonen-Seppanen S, Kirjavainen M, Tammi M, Tammi R, Urtti A. Epidermal cell culture model derived from rat keratinocytes with permeability characteristics comparable to human cadaver skin. Eur J Pharm Sci 2003; 20(1):107–113.

30. Monteiro-Riviere NA, Baynes RE, Riviere JE. Pyridostigmine bromide modulates topical irritant-induced cytokine release from human epidermal keratinocytes and isolated perfused porcine skin. Toxicology 2003; 83(1–3):15–28.

31. Grafe F, Wohlrab W, Neubert R, Brandsch M. Carrier-mediated transport of clonidine in human keratinocytes. Eur J Pharm Sci 2004; 21:309–312.

32. Berner B. Pharmacokinetics of transdermal drug delivery. J Pharm Sci 1985; 74:718–721.

33. Johan Gabrielsson, Daniel Weiner. Pharmacokinetic and Pharmacodynamic Data Analysis: Concepts and Applications. 3rd ed. Stockholm: Swedish Pharmaceutical Press, 2000.

34. Nakayama K, Matsuura H, Asai M, Ogawara K, Higaki K, Kimura T. Estimation of intradermal disposition kinetics of drugs: I. Analysis by compartment model with contralateral tissues. Pharm Res 1999; 16(2):302–308.

35. Auclair B, Sirois G, Ngoc AH, Ducharme MP. Novel pharmacokinetic modelling of transdermal nitroglycerin. Pharm Res 1998; 15(4):614–619.

36. Tashiro Y, Kato Y, Hayakawa E, Ito K. Iontophoretic transdermal delivery of ketoprofen: effect of iontophoresis on drug transfer from skin to cutaneous blood. Biol Pharm Bull 2000; 23(12):1486–1490.

37. Dokoumetzidis A, Macheras P. Investigation of absorption kinetics by the phase plane method. Pharm Res 1998; 15(8):1262–1269.

38. Macpherson SE, Barton CN, Bronaugh RL. Use of in vitro skin penetration data and a physiologically based model to predict in vivo blood levels of benzoic acid. Toxicol Appl Pharmacol 1996; 140(2):436–443.

39. Valiveti S, Hammell DC, Earles DC, Stinchcomb AL. Transdermal delivery of the synthetic cannabinoid WIN 55,212-2: in vitro/in vivo correlation. Pharm Res 2004; 21:1137–1145.

40. Qi X, Liu RR, Sun D, Ackermann C, Hou H. Convolution method to predict drug concentration profiles of 2,3,5,6-tetramethylpyrazine following transdermal application. Int J Pharm 2003; 259:39–45.

41. Luzardo-Alvarez A, Delgado-Charro MB, Blanco-Mendez J. In vivo iontophoretic administration of ropinirole hydrochloride. J Pharm Sci 2003; 92:2441–2448.

42. van der Geest R, van Laar T, Gubbens-Stibbe JM, Bodde HE, Danhof M. Iontophoretic delivery of apomorphine. II: an in vivo study in patients with Parkinson's disease. Pharm Res 1997; 14(12):1804–1810.

43. Riviere JE, Williams PL, Hillman RS, Mishky LM. Quantitative prediction of transdermal iontophoretic delivery of arbutamine in humans with the in vitro isolated perfused porcine skin flap. J Pharm Sci 1992; 81:504–507.
44. Simonsen L, Jorgensen A, Benfeldt E, Groth L. Differentiated in vivo skin penetration of salicylic compounds in hairless rats measured by cutaneous microdialysis. Eur J Pharm Sci 2004; 21:379–388.
45. Cross SE, Anderson C, Roberts MS. Topical penetration of commercial salicylate esters and salts using human isolated skin and clinical microdialysis studies. Br J Clin Pharmacol 1998; 46:29–35.
46. Hashimoto Y, Murakami T, Kumasa C, Higashi Y, Yata N, Takano M. In vivo calibration of microdialysis probe by use of endogenous glucose as an internal recovery marker: measurement of skin distribution of tranilast in rats. J Pharm Pharmacol 1998; 50:621–626.
47. Murakami T, Yoshioka M, Yumoto R, Higashi Y, Shigeki S, Ikuta Y. Topical delivery of keloid therapeutic drug, tranilast, by combined use of oleic and propylene glycol as a penetration enhancer: evaluation by skin microdialysis in rats. J Pharm Pharmacol 1998; 50:49–54.
48. Benfeldt E, Serup J, Menne T. Effect of barrier perturbation on cutaneous salicylic acid penetration in human skin: in vivo pharmacokinetics using microdialysis and non-invasive quantification of barrier function. Br J Dermatol 1999; 140:739–748.
49. Schnetz E, Fartasch M. Microdialysis for the evaluation of penetration through the human skin barrier—a promising tool for future research? Eur J Pharm Sci 2001; 12(3):165–174.
50. Ault JM, Riley CM, Meltzer NM, Lunte CE. Dermal microdialysis sampling in vivo. Pharm Res 1994; 11:1631–1639.
51. Groth L, Serup J. Cutaneous microdialysis in man: effects of needle insertion trauma and anaesthesia on skin perfusion, erythema, and skin thickness. Acta Derm Vereol 1998; 78:5–9.
52. Groth L, Jorgensen A, Serup J. Cutaneous microdialysis in the rat: insertion trauma studied by ultrasound imaging. Acta Derm Vereol 1998; 78:10–14.
53. Mathy FX, Preat V, Verbeeck RK. Validation of subcutaneous microdialysis sampling for pharmacokinetic studies of flurbiprofen in the rat. J Pharm Sci 2001; 90:1897–1906.
54. Chaturvedula A, Joshi DP, Anderson C, Morris R, Sembrowich WL, Banga AK. Dermal, subdermal, and systemic concentrations of granisetron by iontophoretic delivery. Pharm Res 2005; 22(8):1313–1319. Epub 2005 Aug 3.

9

In Vitro–In Vivo Correlation: A Regulatory Perspective with Case Studies

Patrick J. Marroum

Office of Clinical Pharmacology and Biopharmaceutics, Center for Drug Evaluation and Research, Food and Drug Administration, Silver Spring, Maryland, U.S.A.

INTRODUCTION

The recent advances in dissolution methodologies coupled with the availability of sophisticated modeling software enabled dissolution testing to be used more and more both by the industry and regulatory agencies, as a predictor of differences in bioavailability. When drug release from the formulation and its solubilization are the rate limiting steps, it is possible to predict the resulting plasma concentration time profile from its in vitro dissolution. In order to achieve this, there should be a well-established relationship between the in vitro dissolution of the drug from the formulation and its in vivo bioavailability.

This chapter will give an overview for the various requirements from a regulatory perspective for establishing, validating, and applying an in vitro–in vivo correlation (IVIVC). Case studies will be presented on how a predictive IVIVC could be used from both a regulatory and industrial perspective to obtain in vivo bioavailability waivers, as well as setting meaningful dissolution specifications. These case studies will encompass both conventional modified release (MR) dosage forms and specialized dosage forms such as implantables and vaginal rings. Finally issues relating to the elution of drug-eluting stents (DES) will be briefly discussed.

LEVELS OF CORRELATIONS

Level A Correlation

A Level A correlation is a point to point relationship between in vitro dissolution and the in vivo input rate, as can be seen in Figure 1. Such relationships are usually linear where the in vitro dissolution and the in vivo input curves can be superimposable. Even though non-linear relationships are uncommon, they can be appropriate since they are useful in predicting the plasma concentration time profile from in vitro dissolution data (1).

Level B Correlation

In a Level B correlation, the mean in vitro dissolution time is compared either to the mean residence time or the mean in vivo dissolution time (Fig. 2). A Level B IVIVC uses the principles of statistical moment analysis. Even though a Level B correlation uses all the in vitro and in vivo data, it is not considered a point-to-point correlation. It does not uniquely reflect the actual plasma

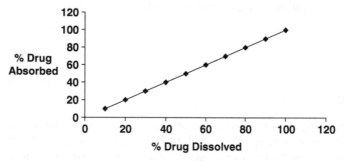

Figure 1 Level A correlation showing the point-to-point relationship between the fraction of drug absorbed and the fraction of drug dissolved.

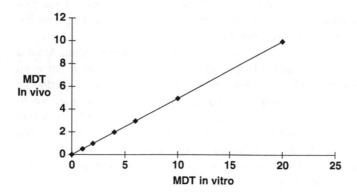

Figure 2 Level B correlation showing the relationship between the mean in vitro dissolution and the mean in vivo dissolution time.

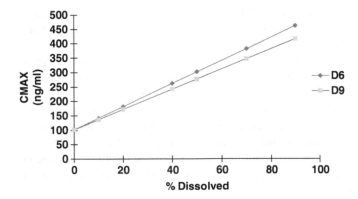

Figure 3 Level C correlation showing the relationship between the amount of drug dissolution at a certain time (e.g., six and nine hours) and the peak plasma concentration.

concentration time profile, because a number of different in vivo profiles will produce similar mean residence times. For this reason, a Level B correlation is of little value from a regulatory point of view.

Level C Correlation

A Level C correlation establishes a relationship between a dissolution parameter, such as the amount of drug dissolved at a certain time, and a pharmacokinetic (PK) parameter of interest such as area under the concentration time curve (AUC) or peak plasma concentration (C_{max}). Unfortunately, a Level C IVIVC does not reflect the complete shape of the plasma concentration time profile, which is a critical factor in defining the performance of the product. On the other hand, a multiple Level C correlation (Fig. 3) relates one or several PK parameters to the amount of drug dissolved, at several time points of the dissolution profile. In general, if one is able to establish a multiple Level C correlation then a Level A correlation could also be established and is the preferred correlation to establish.

DEVELOPMENT OF A LEVEL A CORRELATION

In Vivo Considerations

Since a Level A correlation is the most useful IVIVC both from a regulatory and formulation development point of view, the development of a Level A IVIVC will only be discussed in this chapter.

The following points should be taken into consideration when developing an IVIVC: Sine the PK properties of a drug tend to be somewhat different in animals than humans, only human data is considered from a regulatory point of view. This does not preclude the use of animal data in assessing the perform-ance of pilot formulations.

The in vivo PK studies should be large enough to characterize adequately the product under study. In general, the larger the variability in the performance of the formulation the bigger the study should be (2).

The preferred study design is crossover, since this design reduces inter-subject and inter-study variability, parallel studies, as well as data obtained across several studies can be utilized to develop the IVIVC.

Inclusion of an immediate release reference in the studies will facilitate the data analysis, since it will allow to better estimate the terminal rate constant for each subject and will also enable one to normalize the data to a common reference. The reference product could be an intravenous solution, an aqueous oral solution, or an immediate release product. The earlier described process is illustrated in Figure 4.

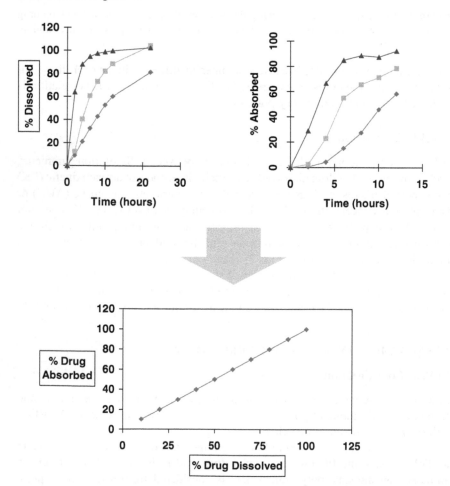

Figure 4 Development of a Level A correlation where the fraction of drug dissolved at each time is plotted against the corresponding fraction of drug absorbed at the same time.

The studies are usually conducted under fasting conditions. However, if there are any tolerability concerns, the studies could be conducted under fed conditions.

In Vitro Considerations

Any in vitro dissolution method may be used to characterize the dissolution behavior of the MR product. The most common dissolution apparatus is USP apparatus I (basket) or II (paddle), at compendially recognized rotation speeds (100 rpm for the basket and 50–75 rpm for the paddle). An aqueous medium with a pH not exceeding 6.8 is recommended as the initial medium for IVIVC development. For poorly soluble drugs, addition of a surfactant such as sodium lauryl sulfate may be appropriate. Nonaqueous or hydralcoholic systems are generally discouraged unless all attempts with aqueous media failed to provide a meaningful dissolution profile. The dissolution profiles should be generated with at least 12 individual dosage units from each lot. A suitable dissolution method should not result in more than a 10% coefficient of variability for the tested batch.

The in vitro dissolution methodology should adequately discriminate among formulations. Dissolution testing can be carried out initially using several methods. Once a discriminating method is chosen, the dissolution method and conditions should be the same for the formulations tested in the biostudy.

Methodology

The IVIVC should be usually developed with two or more formulations (preferably three formulations) with different release rates. The process involves the following steps:

1. In vitro dissolution profiles should be generated using an appropriate dissolution methodology than can discriminate among the various formulations.
2. Determine the plasma concentration time profiles for the tested formulations.
3. Obtain the absorption time profile for these formulations (fraction of drug absorbed vs. time). This can be achieved by the use of appropriate deconvolution techniques.
4. The in vivo absorption profile or the in vivo dissolution profile is plotted against the in vitro dissolution profile to determine whether a relationship exists or not.

The earlier described method is called a two-stage procedure (3). An alternative approach is based on a convolution procedure that attempts to model the relationship between in vitro dissolution and plasma concentrations in a single step. The model predicted plasma concentrations are directly compared to the actual plasma concentrations obtained in the studies (4,5).

EVALUATION OF THE PREDICTABILITY OF THE IN VITRO–IN VIVO CORRELATION

Once an IVIVC has been established, a crucial determination of its applicability is its ability to predict the plasma concentration time profile accurately and consistently.

A relationship between the in vitro dissolution and the in vivo absorption that is dependent on the release rate of the formulation, as can be seen in Figure 5, is usually an indication that a consistent relationship predictive of the in vivo performance does not exist. This is due to the fact that depending on the formulation used, one can have different amount of drug absorbed for the same amount of drug dissolved. On the other hand, a good and consistent relationship would always give you approximately the same slope irrespective of the formulation (whether a slow, fast, or medium formulation is used and whether all the data is pooled together). A good illustration of a valid linear Level A correlation is presented in Figure 6, where the slope of the relationship is the same for each of the individual formulation, or for the case where all the formulations are pooled together and treated as one.

Since the IVIVC model is going to be used to predict the plasma concentration time profile, it is, therefore, imperative to assess the predictive performance of the model via the assessment of its prediction error. Depending on the intended application of the IVIVC and the therapeutic index of

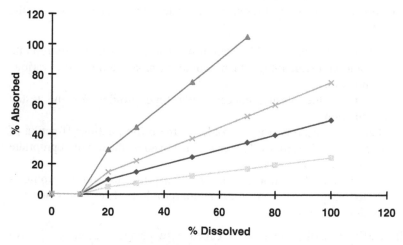

Figure 5 Predictive in vivo–in vitro correlation independent of the release rate where the slope of the relationship is independent of the formulation used.

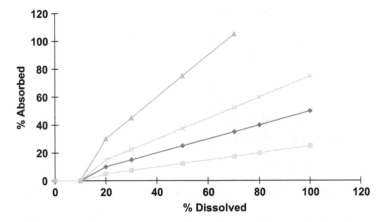

Figure 6 Poor in vivo–in vitro correlation, where the slope of the relationship is dependent on the formulation.

the drug, evaluation of the internal or external predictability may be warranted. Evaluation of internal predictability is based on the data that was used to develop the IVIVC. Evaluation of external predictability involves additional data sets that were not used in the initial development of the IVIVC (6).

If the IVIVC for a non-narrow therapeutic drug was developed with formulations with three or more release rates, the evaluation of the internal predictability would be sufficient to determine its acceptability.

External predictability is warranted in the following situations:

- The drug is considered to be a narrow therapeutic drug.
- The internal predictability criteria are not met.
- The IVIVC was developed with two formulations with different release rates.

The data set that is used in the external predictability should ideally be obtained from a formulation with a different release rate; it is acceptable to use formulations with similar release rates as those used in the development of the IVIVC. The following represent in decreasing order of preference the types of formulations that can be used to estimate the external prediction errors:

- Formulations with different release rates.
- A formulation that was made involving a certain manufacturing change (equipment, process, site etc.).
- Similar formulations but different lots than the ones used in the IVIVC and the data from a different study than the one used in the development of the IVIVC.

APPROACHES TO THE EVALUATION OF PREDICTABILITY

The most common approach to evaluating the predictability of an IVIVC is depicted in Figure 7. The procedure involves the conversion of the in vitro dissolution rate into in vivo absorption rate and then by the use of convolution methods predict the plasma concentration time profile. The AUC and the C_{max} from the predicted profiles are compared to those obtained from the observed profiles to calculate the % prediction errors.

The absolute prediction errors are calculated as follows:

$$[(\text{Observed} - \text{Predicted})/\text{Observed}]^*100$$

These calculations should be done for each of the formulations used to develop the IVIVC.

For internal predictability an average absolute prediction of less than 10% for both AUC and C_{max} establishes the predictive ability of the IVIVC. In addition, the % error for each formulation should not exceed 15%. If the above criteria are not met, the IVIVC is declared inconclusive, and in this case, the evaluation of the external predictability of the IVIVC is required.

For external predictability, the % prediction error should be less than 10% to declare the IVIVC acceptable. A % prediction error between 10% and 20% is

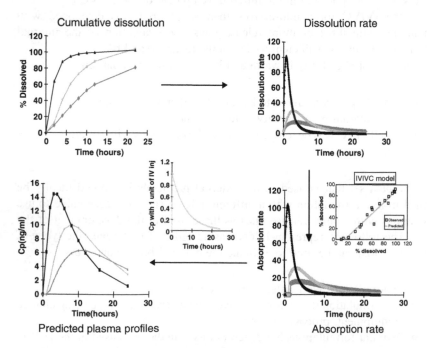

Figure 7 Most common approach in evaluating the predictability of an in vivo–in vitro correlation.

deemed inconclusive, requiring the further evaluation with additional data sets. A percentage prediction error greater than 20% indicates that the IVIVC has a poor predictive ability and thus considered not useful for any application.

It is to be noted that the prediction should be made using mean data (mean dissolution profiles as well as population means for the PK parameters) for the following reasons:

- Individual dissolution data on the dosage unit where the individual subject was administered is not available. Thus, using average in vitro parameters and individual PK parameters is not appropriate.
- Since the purpose of the IVIVC is to predict the performance of yet untested formulations, no individual data will be available for such formulations and, therefore, decision as to the appropriateness of the in vivo performance of the formulations is best determined on the average performance of these formulations (2).

APPLICATIONS OF IN VITRO–IN VIVO CORRELATION

In Vivo Bioavailability Waivers

With a predictive IVIVC, in vitro dissolution would not only be a tool to assure the consistent performance of the formulation from lot to lot, but would become a surrogate from the in vivo performance of the drug product. The ability to predict the plasma concentration time profile from in vitro data will reduce the number of studies required to approve and maintain a drug product on the market, therefore, reducing the regulatory burden on the pharmaceutical industry.

Once an IVIVC has been established, it is possible to waive the requirements for bioavailability/bioequivalence studies. For example, a biowaiver can be granted for a Level 3 process change as defined in SUPAC MR, complete removal or replacement of nonrelease-controlling excipient, and Level 3 changes in the release-controlling excipients.

If the IVIVC is developed with the highest strength, waivers for changes made with the lowest strengths are possible if these strengths are compositionally proportional or qualitatively the same, the in vitro dissolution profiles are similar and all the strengths have the same release mechanism (7).

However, an IVIVC cannot be used to gain the approval of (*i*) a new formulation with a different release mechanism, (*ii*) a dosage strength higher or lower than the doses that have been shown to be safe and effective in the clinical trials, (*iii*) another's sponsor oral-controlled release product even with the same release mechanism, and (*iv*) a formulation change involving an excipient that will significantly affect drug absorption.

Criteria for Granting Waivers

The regultory criteria for granting biowaivers is outlined in the Food and Drug Administration (FDA) guidance on this topic. Basically, the mean predicted

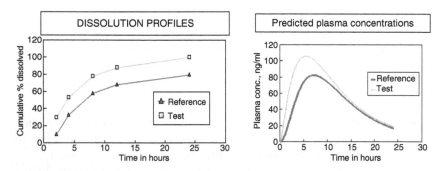

Figure 8 Regulatory criteria for granting a biowaiver using an in vivo–in vitro correlation.

C_{max} and AUC from the respective in vitro dissolution profiles should differ from each other by no more than 20% (Fig. 8) (2).

Dissolution Specifications

The IVIVC allows shifting the dissolution criteria from the in vitro side to the in vivo side. The plasma concentrations time profiles that correspond to the lots that are on the upper and lower limits of the dissolution specifications are predicted. Acceptable dissolution specifications limits are limits that do not result in more than 20% difference in AUC and C_{max} (usually $\pm$ 10% of the target/bioformulation) (see Fig. 9) (8).

Using the IVIVC to choose clinically meaningful specifications provides several advantages in that (*i*) it will minimize the release of lots that are different in their in vivo performance thus optimizing the performance of the product, and (*ii*) in certain cases, will allow wider dissolution specifications.

Release Rate Specifications

The FDA guidance also allows for a novel approach in setting dissolution specifications for formulations exhibiting a zero order release characteristic. An example of such a formulation is the osmotic delivery system, commonly referred to as gastro intestinal therapeutic systems . If these formulations are designed to deliver the drug at a constant rate that can be described by a linear relationship over a certain period of time, then one can set a release rate specification to describe the performance of the formulation.

This release rate specification can be in addition to or instead of the cumulative dissolution specifications that one usually sets for a MR product.

Having a release rate specification will provide for a better control of the in vivo performance of the drug, because it is the release characteristics

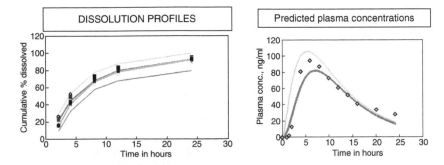

Figure 9 Regulatory criteria for setting the dissolution specifications based on a Level A correlation. The in vivo dissolution profiles are converted to dissolution rates which in turn are converted to in vivo dissolution or absorption profiles. The in vivo rate is convolved with the unit impulse function to obtain the corresponding plasma profile.

of the formulation that determines the rate of appearance in the systemic circulation. This can be described more appropriately by the release rate compared to the cumulative amounts of drug dissolved at a certain interval of time.

As an illustration of this point, let's consider the dissolution profiles of two lots of the same formulation (Fig. 10), with similar release rates but are on the upper and lower limits of the cumulative dissolution specifications. Assuming a Level A correlation for this product, the predicted plasma concentration time profile corresponding to these two lots are similar, differing only in the time to achieve peak plasma concentration. On the other hand, if one examines the case presented in Figure 11, whereby the two lots are very close in their cumulative dissolution profiles (both at the upper limit of the dissolution specifications) but markedly different in their release rates, one can clearly see that the predicted plasma profiles corresponding to these lots are very different and considered not to be bioequivalent (7).

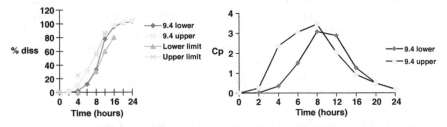

Figure 10 Formulations with similar release rates but on the upper and lower limit of the cumulative dissolution specifications with their corresponding plasma concentration time profiles.

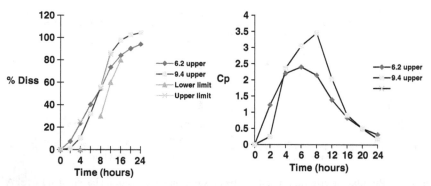

Figure 11 Formulations with different release rates but are on the upper limit of the cumulative dissolution specifications with their corresponding plasma concentration time profiles.

EXAMPLES

Application of a Level A Correlation from a Regulatory Perspective

Toprol XL is a once-a-day formulation of metoprolol approved for the treatment of hypertension and angina. The Level A correlation was obtained from a five-way crossover study with ten subjects, where each subject was given a slow (3.3 %/hr), medium (5.7 %/hr), and a fast (7 %/hr) formulation. As a reference, each subject was also given an oral solution as well as an intravenous infusion of the drug. From the above data, the sponsor was able to develop a Level A correlation between the amount of drug dissolved in vitro and the corresponding amount absorbed in vivo. This correlation was used to grant the sponsor an in vivo bioavailability/bioequivalence waiver for a major formulation change, involving the removal of one of the inert excipients. In view of the fact that both the old and new formulation had almost super-imposable dissolution profiles under the conditions where a predictable IVIVC exists, a waiver for the bioequivalence study was granted. Subsequently, the sponsor changed the manufacturing process to switch from an organic solvent to an aqueous solvent, for which they went ahead and performed a bioequivalence study. Figure 12 shows the plasma concentrations of the old formulation made with the old formulation as compared to the new formulation made with the new process. Table 1 shows the corresponding PK parameters meeting the required regulatory criteria for bioequivalence.

The study was a validation of the Agency's thinking during this time that one could really predict the in vivo performance of a MR formulation from its in vitro dissolution (9).

In Vitro–In Vivo Correlation for a Specialized Dosage Form

NuvaRing is a novel combined contraceptive vaginal ring containing 11.7 mg of etonogestrel (ENG) and 2.7 mg of ethinyl estradiol (EE). It is designed to

Results (Continued):

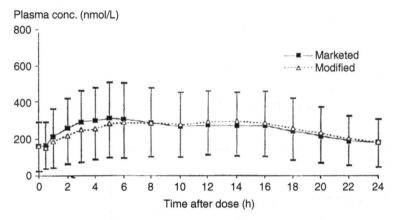

Figure 12   Steady-state plasma concentration time profile for the old and new formulations.

nominally release 120 μg of ENG and 15 μg of EE for 21 days. As part of the development of the ring, the sponsor conducted various dissolution studies to investigate the effect of various conditions, varying in stirring speed as well as pH and ionic strength along with different concentrations of surfactant. The results of these studies show that the release of drug from the NuvaRing is independent from the testing conditions used.

Moreover, during the different stages of development, PK studies were performed with the Silastic prototypes with different ENG loading and release rates, in order to investigate whether a relationship exists between in vitro release rates and in vivo absorption rates. Figure 13 shows the in vitro release rate for the various prototypes tested with the corresponding corrected

Table 1   Bioequivalence Parameters Comparing the Old and New Formulations

	Modified formulation	Original formulation	90% CI
AUC	6129	6073	98–108
C_{max}	316	327	93–103
C_{min}	160.7	165	
T_{max}	10.1	6.14	
FIss	0.67	0.73	

Abbreviations: AUC, area under the concentration time curve; C_{max}, peak plasma concentration; CI, confidence interval; FIss, fluctuation index at steady state.

ENG serum concentrations after five days of NuvaRing treatment corrected for inter-study variability. Figure 14 shows the linear relationship between the in vitro release rate and the in vivo absorption rates for both ENG and EE. The sponsor upon the request from FDA submitted data that validate the IVIVC both internally and externally. Figure 15 shows the predicted versus observed plasma concentration time profile for both ENG and EE, whereas Figure 16 shows the external predictability data for both the components of the ring. The results show that the internal predictability criteria were met for both the

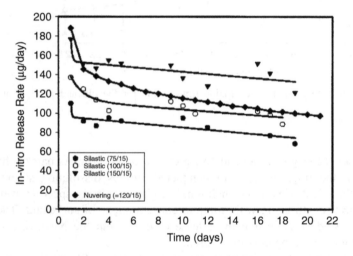

Serum ENG concentrations from the Silastic® prototype rings were corrected in the following figure:

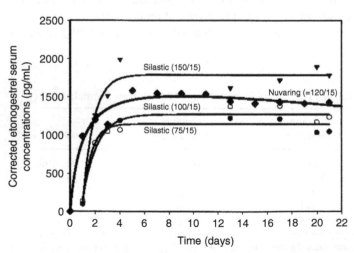

Figure 13 Plasma concentration time profiles for the different prototype rings.

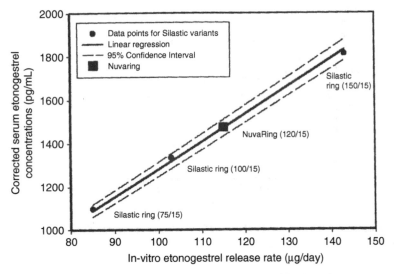

Note: The serum ENG (etonogestrel) concentrations for the Silastic variants were corrected for interassay variation.

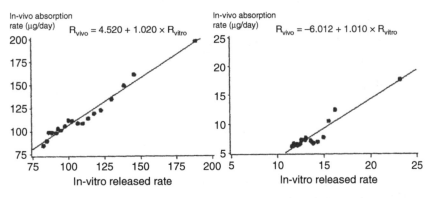

Figure 14 In vivo—in vitro correlation showing the relationship between release rate and plasma concentrations.

components (10). Subsequently, the sponsor changed the batch size, the manufacturing process, and the amount of noncontrolling excipients. Changes for such a product ordinarily would require a bioequivalence study. However, since the IVIVC was deemed acceptable and predictive of in vivo performance, the sponsor was able to obtain an in vivo bioequivalence waivers based on the comparability of dissolution profiles.

Developing an IVIVC for such specialized dosage forms where the conduct of bioequivalence is not as easy as conventional dosage forms is very desirable, because it reduces the regulatory burden from having to conduct many in vivo

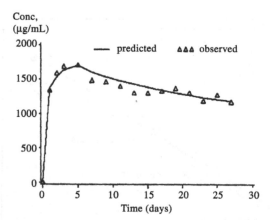

Internal validation: Predicted *versus* observed in-vivo serum concentrations for etonogestral

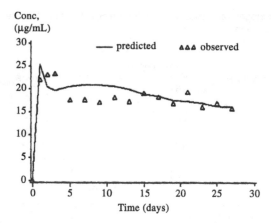

Internal validation: Predicted *versus* observed in-vivo serum concentrations for ethinylestradiol for 1

Figure 15 Internal predictability showing the observed versus predicted plasma concentration time profiles.

studies whenever a manufacturing change is affected to the product as could be seen from the previous example (11).

Drug-Eluting Stents

With the recent advances in medical technology, it is more common to see the therapeutic effect of a device be optimized by its combination with a drug. A prime example of such a device is the DES. Since these stents are implanted,

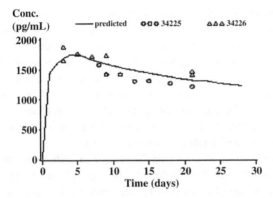

External validation: Predicted vs observed in vivo concentrations
for etonogestrel

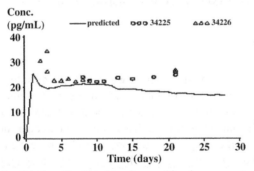

External validation: Predicted vs observed in vivo concentrations
for ethinylestradiol

Figure 16 External predictability showing the observed versus predicted plasma concentration time profiles.

having consistent elution characteristics throughout the intended duration of action is crucial in maintaining the therapeutic benefit to the patient. Due to the extreme difficulty in estimating the in vivo elution characteristics for such devices, setting elution specifications that will be relevant from an in vivo point of view becomes very challenging.

In the case where the measurable plasma levels are indicative of the in vivo elution of the drug from the stent at the site of action and the in vitro conditions result in in vitro elution rates mimicking those observed in vivo, the dissolution specifications should be set in terms of the observed in vitro elution rate.

However, in the situation where the plasma levels are too low to measure, it becomes practically impossible to determine the elution characteristics. In such a case, animal models could be used to determine the elution characteristics of the

DES. At different time intervals, the stents could be explanted and the amount of drug remaining on the stent as well as the amount found in the adjacent tissues could be measured. This information can be a valuable guide for the development of the most relevant elution method with the most relevant specifications. In other situations, with the current advances in X-ray computer technologies, it may be possible to noninvasively monitor the local drug release from the DES. Such a capability will go a long way in characterizing the elution behavior in the target population. This will in turn enable one to select the elution method and specifications with the in vivo considerations in mind (12,13).

Another important consideration in setting the elution specifications is the clinical performance of the DES. If the clinical trials showed that there is a correlation between the safety and efficacy profile and elution rates, the specifications should be set in such a way that only DES with elution rates with acceptable safety and efficacy profiles be released to the market. At a minimum, the elution specifications should not release any lots with elution characteristics beyond what was found to be acceptable from a clinical point of view.

CONCLUSION

The establishment of a predictive relationship between in vitro dissolution and the in vivo bioavailability of a MR formulation will decrease the number of studies required to approve and maintain a product on the drug market. A predictive Level A correlation will enable the in vitro dissolution to become a surrogate for the in vivo performance of the drug product. Consequently, more meaningful dissolution specifications that take into account the in vivo consequences could be established and wider dissolution specifications could be justified based on the predicted outcomes. Thus, a meaningful and predictive correlation can be a useful tool both in product development and regulatory decision making.

REFERENCES

1. US Pharmacopeia 23. General information/in vitro in vivo evaluation of dosage forms. <1088>, 1995:1924–1929.
2. FDA Guidance for Industry. Extended release oral dosage forms: development, evaluation and applications of in vitro/in vivo correlations, Sep 1997.
3. Marroum PJ. Development of in vivo in vitro correlations, SUPAC-MR/IVIVC guidance FDA training program manual. Center for Drug Evaluation and Research, Food and Drug Administration, July 1997:62–76.
4. Gillespie WR. Convolution –Based approaches for in vivo in vitro correlation modeling, in vitro in vivo correlations. Adv Exp Medi Biol 1997; 423:53–65.
5. Gillespie WR. Modeling strategies for in vivo in vitro correlations. In: Amidon G, Robinson JR, Williams RL, eds. Scientific Foundations for Regulating Drug Product Quality. Alexandria, VA: AAPS Press, 1997:275–292.

6. Uppoor R. Evaluation of predictability of in vivo in vitro correlations, SUPAC-MR/ IVIVC guidance FDA training program manual. Center for Drug Evaluation and Research, Food and Drug Administration, July 1997:97–110.
7. Marroum PJ. Regulatory examples: dissolution specifications and bioequivalence product standards. In: Amidon G, Robinson JR, Williams RL, eds. Scientific Foundations for Regulating Drug Product Quality. Alexandria, VA: AAPS Press, 1997:305—319.
8. Marroum PJ. Role of in vivo-in vitro correlations in setting dissolution specifications. Am Pharm Rev 1999; 2:39—42.
9. Cassella P. Clinical pharmacology review and biopharmaceutics review. Center for Drug Evaluation and Research, Food and Drug Administration, April 1991.
10. Lau S. Clinical pharmacology review and biopharmaceutics review. Center for Drug Evaluation and Research, Food and Drug Administration, December 2000.
11. Jaragula V. Clinical pharmacology review and biopharmaceutics review. Center for Drug Evaluation and Research, Food and Drug Administration, March 2002.
12. Szymanski-Exner A, Stowe WT, Salem K, Lazebnik R, Haaga JR, Wilson DL, et al. Noninvasive monitoring of local drug release using x-ray computed tomography: optimization and in-vitro/in-vivo valiation. J Pharm Sci 2003; 92:289.
13. Hwang CW, Wu D, Edelman ER. Physiological transport forces govern drug distribution for stent-based delivery. Circulation 2001; 104:600.

Index